Ayam Cemani C.

The Indonesian Black Hen

A Complete Owner's Guide to this rare pure black chicken breed.

Covering History, Buying, Housing, Feeding, Health, Breeding & Showing.

By Angela Jewitt

Copyright © August 2015

Ayam Cemani Chickens.

All rights reserved. No part of this publication may be reproduced, distributed, or transmitted in any form or by any means, including photocopying, recording, or other electronic or mechanical methods, without the prior written permission of the publisher, except in the case of brief quotations embodied in critical reviews and certain other noncommercial uses permitted by copyright law. For permission requests, write to the publisher, addressed "Attention: Permissions Coordinator," at the address below:

Published by: WHYBANK PUBLISHING, TD1 1UF, UK

ISBN: 978-0-9930278-4-0

Disclaimer

Although the author and publisher have made every effort to ensure that the information in this book was correct at press time, the author and publisher do not assume and hereby disclaim any liability to any party for any loss, injury, damage or disruption caused by errors or omissions, whether such errors or omissions result from negligence, accident, non-functional websites, or any other cause. Any advice or strategy contained herein may not be suitable for every individual.

Foreword

If you are thinking about purchasing Ayam Cemani chickens, this book is the perfect place to start. Not only will you find an introduction to the breed as well as some Ayam Cemani facts, but you will also find a wealth of information about caring for, feeding, and breeding these beautiful chickens. By the time you finish this book you will know for certain whether the Ayam Cemani is the right breed for you and, if it is, you will be well on your way to becoming an Ayam Cemani owner.

Acknowledgements

I would like to extend my sincerest thanks to my friends and family for their support as I wrote this book.

Special thanks goes to Kathryn at Homes4Hens commercial hen rescue in Dumfries and Galloway for inspiring me to keep hens in the first place. She showed us what amazing animals they are to keep even though they have such a terrible and unnecessarily cruel start to life.

More special thanks to my two beautiful children for all the egg collecting.

And finally, thanks to Fergus at Peggys Pekins in Norfolk for allowing me to use his fantastic photographs.

Table of Contents

Chapter One: Introduction...1

Useful Terms to Know..3

Chapter Two: Understanding Ayam Cemani Chickens.......6

1.) What Are Ayam Cemani Chickens?.....................7

2.) Facts About Ayam Cemani Chickens 9

Summary of Facts...10

3.) History of Ayam Cemani Chickens as Pets.........12

4.) Ayam Cemani Chicken Varieties.......................14

5.) Ayam Cemani Chickens as Food......................17

Chapter Three: What to Know Before You Buy.............18

1.) Do You Need a Licence? 19

a.) Licencing in the U.S. 19

b.) Licencing in the U.K. 20

2.) How Many Should You Buy?...........................22

3.) Can Ayam Cemani Chickens Be Kept with Other Pets?..24

4.) Ease and Cost of Care...................................26

a.) Initial Costs 26

b.) Monthly Costs 28

5.) Pros and Cons of Ayam Cemani Chickens 32

Chapter Four: Purchasing Ayam Cemani 33

1.) Where to Buy Ayam Cemani Chickens 34

a.) Buying in the U.S. 34

b.) Buying in the U.K. 36

2.) How to Select a Healthy Ayam Cemani Chicken 38

Chapter Five: Caring for Ayam Cemani 41

1.) Habitat Requirements 42

a.) Space Requirements for Chickens 43

b.) Types of Chicken Coops 44

c.) Raising Chickens in Urban Areas 48

2.) Building a Chicken Coop 50

3.) Maintaining Your Chicken Coop 53

4.) Keeping Your Chickens Warm 54

Chapter Six: Feeding Ayam Cemani Chickens 56

1.) Nutritional Needs of Chickens 57

2.) Types of Chicken Feed 60

3.) Other Types of Food 63

4.) Ayam Cemani Feeding Recommendations 66

Chapter Seven: Breeding Ayam Cemani 67

1.) Basic Breeding Info 68

2.) The Breeding Process 70

3.) Hatching Ayam Cemani Eggs 72

4.) How to Deal with Culling 74

Chapter Eight: Keeping Ayam Cemani Chickens Healthy 78

1.) Common Health Problems 79

2.) Preventing Illness 89

Chapter Nine: Showing Ayam Cemani 91

  1.) Breed Standard 92

  2.) What to Know Before Showing...........................96

Chapter Ten: Ayam Cemani Care Sheet 98

  1.) Basic Information 99

  2.) Habitat Set-up Guide.......................................100

  3.) Nutritional Information 102

  4.) Breeding Tips...103

Chapter Eleven: Relevant Websites...................................105

  1.) Food for Ayam Cemani Chickens 106

  2.) Coops and Coop Supplies.................................108

  3.) Breeding and Incubation Supplies 110

  4.) General Info for Raising Chickens.....................112

Index...115

Photo Credits 123

References..125

Chapter One: Introduction

Chapter One: Introduction

When you think of a chicken, what do you picture? A big white rooster with a bright red crown? Or perhaps you think of a brown hen with no crown at all. The Ayam Cemani chicken is not your average chicken – it is completely black. Not only does the Ayam Cemani have pitch-black feathers but its skin, meat, and even its internal organs and bones are completely black. No other chicken is quite like the Ayam Cemani.

The Ayam Cemani is a very rare and unique breed of chicken that comes from the island of Java, Indonesia. This breed of chicken is known not only for its unique black

Chapter One: Introduction

coloration but for its prominence in Indonesian religious and mystical practices. The meat of the Ayam Cemani is said to have magical properties and this bird has been used in religious sacrifices for hundreds of years. Though the Ayam Cemani is still a rare breed, it is gaining in popularity – especially since being introduced in Europe and the United States in the late 1990s.

If you are thinking about raising Ayam Cemani chickens yourself, you would be wise to learn as much as you can about them before you begin. The Ayam Cemani is different from the average chicken in a number of ways and it requires certain conditions and nutrients to remain healthy. In this book you will find a wealth of information about the Ayam Cemani including facts about the breed, its history, and tips for care. You will also receive a guide for building your own chicken coop, for breeding Ayam Cemani, and for keeping your birds healthy. By the time you finish this book you will know for certain whether the Ayam Cemani is the right breed for you and, if it is, you will be well on your way to becoming an Ayam Cemani owner.

Chapter One: Introduction

Useful Terms to Know

American Poultry Association (APA) - The organization in charge of setting the standard for purebred poultry in the United States.

Beak – The hard, pointed protrusion from the face of the chicken, forming the nose and mouth.

Beard – A group of feathers below the bird's beak.

Bedding – The material spread on the floor of a chicken coop or brooding area; examples include straw, hay, wood shavings, alfalfa grass, etc.

Biddy – A slang term for hen.

Broiler – A young chicken, usually under 12 weeks old, bred specifically for meat.

Brood – To care for baby chickens; also a name for a group of baby chicks.

Brooder – A heated enclosure that is used to raise chicks.

Clutch – A group of eggs laid at the same time.

Cock – A male chicken; also a rooster.

Cockerel – A male chicken under 1 year old.

Comb – A fleshy outcrop of skin on the top of a chicken's head; used to help regulate body temperature.

Coop – A structure used to house chickens.

Ayam Cemani Chickens

Chapter One: Introduction

Down – The soft fur-like feathers covering a newly hatched chick.

Dusting – A method of cleaning in which a chicken rolls in the dust or dirt.

Flock – A group of poultry.

Fowl – Domestic birds that are typically raised for food.

Free-Range – Used to describe chickens or other fowl that are allowed to roam a pasture at will.

Fryer – A young chicken with tender meat.

Hackles – The feathers on a rooster's cape.

Hatchability – The percentage of fertilized eggs that hatch in an incubator.

Hen – A female chicken.

Incubate – To establish and maintain hatchable conditions for fertilized eggs.

Incubation Period – The period of time during which fertile eggs are incubated in order to hatch.

Layer – A female chicken that has started laying eggs.

Pecking Order – A social ranking among chickens.

Plumage – The feathers on a bird.

Pullet – A female chicken less than 1 year old.

Purebred – A bird that comes from a rooster and hen belonging to the same species.

Roaster – A pullet or cockerel suitable for cooking whole.

Chapter One: Introduction

Set – To sit on eggs to keep them warm; to brood.

Standard – A description of the ideal characteristics of a particular breed.

Starter Feed – The type of feed given to newly hatched poultry with a higher protein concentration to support growth and development.

Vent – The opening of the cloaca through which eggs and excrement are passed.

Chapter Two: Understanding Ayam Cemani Chickens

Chapter Two: Understanding Ayam Cemani Chickens

Before you can decide whether or not the Ayam Cemani chicken is the right breed for you, you need to learn as much as you can about them. Where do these chickens come from? What makes them black? Why are they so rare? In this chapter you will learn the basics about the Ayam Cemani chicken as well as some facts about its history as pets and the different varieties that exist. You will also learn about the Ayam Cemani chicken as a food source.

Chapter Two: Understanding Ayam Cemani Chickens

1.) What Are Ayam Cemani Chickens?

As you already learned in the introduction to this book, the Ayam Cemani chicken is an incredibly rare and unique bird. This species of chicken comes from the island of Java off the coast of Indonesia and it is known for its black coloration. While some chicken breeds might have a few black feathers or an overall dark coloration, the Ayam Cemani is unique in the fact that it is pitch black over its entire body. The black coloration of this breed is not just limited to its outward appearance – its bones, tissues, and organs are black as well.

Some have referred to the Ayam Cemani as the "Lamborghini of the chicken world" because it is so beautiful and so rare. These chickens have long been valued for their so-called "magical" properties in their native land of Indonesia and they have been known to sell for as much as $2,500 (£2,250) per bird – yes, a single chicken can sell for over $2,000 (£1,800). Not only are these birds valued for their so-called magical properties, but their jet black feathers are something of a commodity as well. The feathers of the Ayam Cemani may be black but they have a metallic sheen with hints of green and purple, much like a beetle.

Though Ayam Cemani chickens are native to Indonesia, stocks of these birds are kept throughout the world in

Chapter Two: Understanding Ayam Cemani Chickens

countries like Germany, the Netherlands, Slovakia, and the Czech Republic. The Ayam Cemani breed has even been introduced into Europe and the United States where breeders can make quite a profit breeding and selling these birds. The tricky thing about this species is that they do not breed as prolifically as other species and they do not roost – the only way to hatch the eggs of this species is to incubate them. This may account for part of the high price this breed has been known to fetch.

Chapter Two: Understanding Ayam Cemani Chickens

2.) Facts About Ayam Cemani Chickens

The Ayam Cemani chicken is a rare and beautiful bird known for its completely black coloration. These birds are average-sized when compared to many domesticated chicken breeds with males of the species growing between 2 and 3.5 kg (4.4 to 7.7 lbs.) and females growing between 1.5 and 2.5 kg (3.3 to 5.5 lbs.). According to the Ayam Cemani standard of perfection, these birds should have a long, flat head with a black beak measuring about 1.8 cm (0.7 inches) long and between 0.9 and 1.2 cm (0.35 to 0.47 inches) wide. The beak is completely black, as is the tongue.

This breed of chicken has a single black comb with three, five or seven points and a coupled smooth wattle. The eyes are round and black, and the neck is medium sized. The male Ayam Cemani exhibits a large, wide chest measuring about 12.5 cm (4.92 inches) in length and about 34.1 cm (13.42 inches) in diameter. The body of the male slants back from the neck while the female's body has a more rectangular, rhombus-like shape. Males have a thin back with strong, sturdy wings while females have a flat, wide back and a flat wingspan. The tail of the Ayam Cemani chicken is scooped like a horse's tail, made completely of black feathers. The claws of the Ayam Cemani are sharp, measuring about 10 cm (3.94 inches) long in males and 6 cm (2.36 inches) long in females.

Chapter Two: Understanding Ayam Cemani Chickens

The coloration of the Ayam Cemani chicken is due to a dominant gene that causes hyperpigmentation – this is referred to as Fibromelanosis. This hyperpigmentation affects all parts of the chicken including the skin, feathers, and even the internal organs. In fact, the name of this species is made in reference to its color. The word *Ayam* means "chicken" in Indonesian while *Cemani* means "completely black" in Javanese. The same hyperpigmentation that affects they Ayam Cemani breed is also seen in other domesticated chicken species like the Swedish Black Hen.

Ayam Cemani chickens are very rare, partially due to the fact that they do not breed as readily as other domesticated species. The Ayam Cemani hen usually only lays about 60 to 100 eggs during their first year with the average cycle producing 20 to 30 eggs. These birds usually stop laying for about 3 to 6 months after each cycle. The eggs laid by the Ayam Cemani chicken are fairly large when compared to the size of the hen's body and they are cream-colored. Each egg weighs about 45 grams at laying. This breed tends to be a poor sitter and these birds do not roost – this means that breeders must incubate the eggs to hatch them.

Summary of Facts

Origins: island of Java, Indonesia

Chapter Two: Understanding Ayam Cemani Chickens

Species: *Gallus gallus domesticus* (domesticated chicken)

Breed Name: Ayam Cemani

Coloration: completely black, inside and out

Genetics: hyperpigmentation caused by a dominant gene (Fibromelanosis)

Weight (male): between 2 and 3.5 kg (4.4 to 7.7 lbs.)

Weight (female): between 1.5 and 2.5 kg (3.3 to 5.5 lbs.)

Beak (length): 1.8 cm (0.7 inches)

Beak (width): between 0.9 and 1.2 cm (0.35 to 0.47 inches)

Comb: single, black; 3, 5 or 7 points

Wattle: couple, smooth, black

Body (male): slants back from the neck; back is thin

Body (female): rectangular, rhombus-like in shape; back is flat and wide

Tail: scooped like a horse's tail

Egg Laying: about 60 to 100 eggs during the first year; 20 to 30 eggs per cycle

Breeding Cycle: hens take a 3 to 6 month break after each cycle

Egg Size: about 45 grams at laying; large compared to the size of the hen

Egg Color: cream-colored

Chapter Two: Understanding Ayam Cemani Chickens

3.) History of Ayam Cemani Chickens as Pets

The exact origins of the Ayam Cemani chicken are unknown, but the black pigment responsible for its color is known to have existed for about 800 years in Asia. Though the exact details of this breed's origin are unknown, scientists have determined that it is one of the seven species belonging to the Indonesian line of jungle fowl. This line of fowl includes a species of black partridge which has the same inside-and-out black coloration of the Ayam Cemani chicken breed.

It is believed that the Ayam Cemani chicken was first described during the 1920s by Dutch colonists. Popular theory suggests that Mr. Tjokromihardjo from a town in central Java called Grabag, Magelang is the so-called "founder" of the breed. Mr. Tjokromihardjo's son is currently the owner of one of the largest commercial layer farms in Java. The Ayam Cemani was introduced in Europe during the late 1990s by Dutch breeders at which point the popularity of the breed skyrocketed. The breed has since been introduced into a variety of other countries including the United States, the Netherlands, Germany, Slovakia, and the Czech Republic.

Though the Ayam Cemani is only starting to become popular outside its native country, it has long been used in

Chapter Two: Understanding Ayam Cemani Chickens

traditional medicine in Indonesia. The meat of the Ayam Cemani is thought to have mystical powers – a sacrifice of the chicken is thought to bring luck. Some Asian cultures believe that eating the meat of the Ayam Cemani chicken will ease an unsettled conscience and that the crowing of the bird brings prosperity. It is also believed that the meat of this chicken is higher in iron, making it beneficial for women both before and after childbirth. Ayam Cemani chickens are sometimes sacrificed during childbirth to bring good fortune to the mother.

Chapter Two: Understanding Ayam Cemani Chickens

4.) Ayam Cemani Chicken Varieties

Some say that the Ayam Cemani is a breed of its own while others maintain that it is simply a color morph of the Ayam Kedu breed. The origins of the Kedu chicken can be traced back to the 1920s when Mr. Tjokromihardjo first recorded it as a separate species. <u>According to Mr. Tjokromihardjo, the Ayam Kedu chicken comes in three different varieties</u>:

- Kedu Kedu
- Kedu Cemani
- Kedu Hsian

The Kedu Kedu is a large chicken variety has black feathers with white skin and a large red or black comb on its head. The Kedu Hsian is a type of partridge with white skin, partridge coloration and a red comb. The Kedu Cemani has black skin, black feathers, and a large black comb. All three varieties grow to about 4 to 6 lbs. (1.8 to 2.7 kg) for hens and 5 to 8 pounds (2.3 to 3.6 kg.) for roosters. The average lifespan of the Ayam Kedu chicken averages between 6 and 8 years. The Kedu Cemani is considered to be a separate breed from the Ayam Kedu in both Europe and in North America.

Another breed of chicken native to Indonesia is the Ayam Pelung, or the Pelung Chicken. This breed is known for its melodious crowing with males of the breed being

Chapter Two: Understanding Ayam Cemani Chickens

nicknamed "singing chickens". A fully grown male Ayam Pelung weights between 12 and 15 pounds (5 to 6 kg) and stands as tall as 18 inches (50 cm). In contrast, hens typically max out around 8 lbs. (3.6 kg). The Ayam Pelung chicken has a sturdy, well-built posture with large blue-black legs and a round red wattle. The coloration of this breed is mixed with no distinctive color – some are a mixture of black and red while others are a mix of yellow, white and green.

The Ayam Pelung chicken has a single comb which is large and red to match the wattle. This breed has a rhythmic crow which is longer and more melodious than most domesticated chickens. This is a factor that contributes greatly to the popularity of the species. It is also important to note that these chickens grow very quickly which makes them a popular choice as broiler chickens and they are generally kept as free-range chickens. Females of the breed start producing eggs early, maxing out at around 70 eggs or so per year.

The Swedish Black Hen, or Svart Höna, is another chicken breed that exhibits the same fibromelanistic trait as the Ayam Cemani. This breed is actually rarer than the Ayam Cemani with an estimated population of only 500 birds existing worldwide. The origins of the Swedish Black Hen are thought to extend back about 400 years ago when the gene causing fibromelanosis found its way into a

Chapter Two: Understanding Ayam Cemani Chickens

population of chickens being transported to Norway. A race of black chickens popped up along the Norway-Sweden border and gave birth to the modern Svart Höna.

Like the Ayam Cemani chicken, the Swedish Black Hen is a fairly small species – roosters weigh only about 4 pounds (1.8 kg) and they have sleek, slender bodies. Females of the species lay cream-colored or white eggs, though the birds themselves are almost entirely black with some deep red facial skin. The feathers of the Swedish Black Hen have a metallic sheen and the birds have a relaxed and friendly temperament. These birds are typically kept as free-range chickens since they are excellent foragers and they can withstand cold temperatures down to 20°F (-6°C) with no negative effects.

Chapter Two: Understanding Ayam Cemani Chickens

5.) Ayam Cemani Chickens as Food

You won't find black chicken meat for sale at your local grocery store, unless you live in certain Asian countries. Due to the high price of these birds, it is difficult to find people who have actually tasted their meat. In fact, Ayam Cemani chickens are largely raised for their red-black blood which is used in ceremonies. The Indonesian people typically do not eat Ayam Cemani chicken meat for meals – they usually only eat it during ceremonies. According to unnamed sources, Ayam Cemani chicken meat is more delicious than the meat of the average chicken but it can be a little gamier since these birds have a fairly narrow carcass.

Chapter Three: What to Know Before You Buy

Now that you know the basics about the Ayam Cemani chicken breed you may have a better idea whether or not it is the right breed for you. Still, you should take the time to learn some of the more practical aspects of keeping these chickens. In this chapter you will learn about the licensing requirements for the Ayam Cemani chicken as well as tips for how many to buy and whether they get along with other animals. You will also receive information about the cost to keep these birds and the pros and cons of the species.

Chapter Three: What to Know Before You Buy

1.) Do You Need a Licence?

Before you go out and buy a flock of Ayam Cemani chickens, you should learn whether or not it is even legal for you to keep them in your area. Certain regions require individuals to obtain a licence to keep poultry and other animals, while others do not. In this section you will receive some basic information about licensing and permit requirements for chickens in the United States and in the United Kingdom.

a.) Licencing in the U.S.

There are no national restrictions against keeping Ayam Cemani chickens, or other poultry, so you will have to learn about individual state, county, and city laws affecting the area in which you live. There are likely to be fewer restrictions on the type and number of poultry you can keep in rural areas than in urban areas, but it is still important that you check with your local council to determine if there are any size or number restrictions. You may also be required to take certain precautions against the spread of disease by registering your flock.

In addition to thinking about restrictions and regulations regarding the keeping of poultry like Ayam Cemani chickens, you should also consider whether there are any

Chapter Three: What to Know Before You Buy

legal protections for these animals. According to the ASPCA there are no national laws protecting farm animals while they are on the farm but there are two laws regarding transport and humane slaughter – unfortunately, both of these laws exclude poultry. Still, you should consult ASPCA literature regarding the humane keeping and slaughtering of poultry like Ayam Cemani chickens.

b.) Licencing in the U.K.

In the United Kingdom there are no national laws that will keep you from legally keeping Ayam Cemani chickens. There are, however, certain regulations you will need to comply with. For example, The Department for Environment Food and Rural Affairs (DEFRA) requires individuals to register their birds if they keep more than 50 on the premises – this includes all types of poultry, both adults and babies. You are not legally required to register your flock if you keep fewer than 50 birds, but you are encouraged to do so.

Even if there are no national rules restricting the keeping of Ayam Cemani chickens, there may be certain bylaws or covenants in place designed to keep people from keeping livestock like chickens. This is generally a local decision so you'll need to check with your local council to see if any of these restrictions affect you. Even if it is legal, however, you

Ayam Cemani Chickens

Chapter Three: What to Know Before You Buy

could still receive complaints from local residents about mess or noise levels so make sure you are respectful of other residents if you choose to keep these birds.

Chapter Three: What to Know Before You Buy

2.) How Many Should You Buy?

When considering how many Ayam Cemani chickens to buy, you need to think about several factors. First, what is your motivation for keeping chickens – are you breeding them for eggs? For meat? Or simply keeping them as pets? Ayam Cemani chickens do not produce as many eggs as other domesticated breeds and they do not tend to sit well - you'll have to incubate the eggs yourself so think about this when considering how many chickens to get.

Second, how much space do you have to devote to your chickens? Ayam Cemani chickens generally do well as free-range birds, though they can tolerate a large outdoor enclosure if necessary. The space you have to devote to your chickens may be limited depending whether you live in a rural or urban area. If you are raising chickens for the eggs, you should plan to keep 5 to 10 hens in order to feed your whole family. When it comes to space, each hen will require a minimum of 4 square feet of indoor space, though they will fare better as free-range birds.

In addition to thinking about how many Ayam Cemani chickens you want to keep, you also need to think about whether you want to keep just hens or roosters as well. Your Ayam Cemani hens will continue to lay eggs whether or not a rooster is present, but they will not be fertile. If you

Chapter Three: What to Know Before You Buy

plan to actually breed and raise your chickens you will need to have at least one rooster and several hens. Keep in mind that chickens are most productive during the first two years, so you'll need to acquire more stock every one or two years if you want to maintain production.

Chapter Three: What to Know Before You Buy

3.) Can Ayam Cemani Chickens Be Kept with Other Pets?

When you keep a group of chickens together they will naturally establish a "pecking order" – a type of social hierarchy. This pecking order determines which chickens get to eat first and which ones have access to the best nest and mates. Because this order is so important in keeping chickens, it can be risky to introduce another species which has its own societal rules for behavior. Avoid keeping Ayam Cemani chickens with very violent types of poultry to prevent problems.

It is also important to consider the transmission of diseases when thinking about whether to keep your chickens with other birds. Chickens are frequent carriers of many different diseases that can be passed on to turkeys, pheasants, and other types of poultry. You would experience a lower risk for disease transmission in keeping your Ayam Cemani chickens with ducks, though there are still some risks. Ducks usually eat different food than chickens so you don't have to worry about competition and they prefer roosts that are lower to the ground. Chickens also use waterers that are too small for ducks to foul by swimming in them. On the other hand, drakes (male ducks) have been known to kill hens and roosters have been known to injure male ducks on occasion.

Chapter Three: What to Know Before You Buy

If you absolutely must keep your Ayam Cemani chickens with another type of bird, consider guinea fowl. Guinea fowl are the most compatible species to keep with chickens, though you do need to be careful about keeping adult chickens with baby guinea fowl. As long as you provide both species with the right food, water, and sleeping quarters you should be fine.

Chapter Three: What to Know Before You Buy

4.) Ease and Cost of Care

In addition to considering the legal issues associated with keeping Ayam Cemani chickens as well as the number of chickens to keep, you also have to think about the costs of keeping these birds. Not only will you need to cover the initial costs of setting up your chicken habitat and buying your chicks, but you will also then need to cover recurring costs for food, bedding, veterinary care, and more. In this section you will receive an overview of both the initial costs and monthly costs for keeping Ayam Cemani chickens.

a.) Initial Costs

The initial costs associated with keeping Ayam Cemani chickens are those costs that you must cover to get started. These include things like the cost of chicks, your chicken coop, feeding and watering stations, and a brooder. You will find an overview of each cost below as well as an estimate cost.

Purchase Price – Because Ayam Cemani chickens are still fairly rare, you may have a difficult time finding them. An adult Ayam Cemani can run up to $2,500 (£2,250) so it would probably be more cost-effective to buy your chickens as chicks and raise them yourself. The average cost for an Ayam Cemani chick is around $200 (£180). You should plan

Chapter Three: What to Know Before You Buy

to keep a minimum of three hens, so plan on a cost around $600 (£540) for your Ayam Cemani chicks.

Chicken Coop – The cost of your chicken coop could vary significantly depending on its size and the materials you choose to use. Even if you plan to let your Ayam Cemani chickens range freely, you should still provide them with a coop so they have some form of shelter overnight and in inclement weather. The cost to build your own chicken coop could be as low as $300 (£270) while the cost to buy a large, prefabricated coop could be over $1,000 (£900).

Feeding/Watering Stations – Just like the chicken coop, your feeding and watering stations vary greatly in cost depending on the size and the materials from which they are made – it also depends how many of them you need according to the number of chickens you plan to keep. You should expect to spend about $50 (£45) for feeding/watering stations for up to three chickens and up to $100 (£90) for a group of 5 or more chickens.

Brooder – If you purchase your Ayam Cemani as chicks you will need to raise them in a brooder. You can buy a prefabricated brooder or make one yourself from a large plastic tub and a heat lamp. The average cost for a brooder is between $75 and $100 (£68 to £90).

Ayam Cemani Chickens

Chapter Three: What to Know Before You Buy

To help you get an idea how much the initial costs for Ayam Cemani chickens are, consult the chart below. You will notice that there are cost estimates for a single chicken and for a group of 3 and a group of 5.

Initial Costs for Ayam Cemani Chickens			
Cost Type	1 Chicken	3 Chickens	5 Chickens
Purchase Price	$200 (£180)	$600 (£540)	$1,000 (£900)
Chicken Coop	$300 - $1,000 (£270 - £900)	$300 - $1,000 (£270 - £900)	$300 - $1,000 (£270 - £900)
Feed/Water Stations	$50 (£45)	$50 (£45)	$100 (£90)
Brooder	$75 to $100 (£68 to $90)	$75 to $100 (£68 to $90)	$75 to $100 (£68 to $90)
Total	$625 - $1,350 (£562-£1,215)	$1,025-$1,750 (£922-£1,575)	$1,475-$2,200 (£1,328-£1,980)

* U.K. costs are estimated based on an exchange rate of $1/£0.90. Rates are subject to change.

b.) Monthly Costs

Once you have purchased your Ayam Cemani chicks and set up your enclosure, you then have to cover recurring costs. Monthly costs for Ayam Cemani chickens include food, bedding, veterinary care and repairs/replacements.

Ayam Cemani Chickens

Chapter Three: What to Know Before You Buy

You will find an overview of each cost below as well as an estimate cost.

Food – When you are raising your Ayam Cemani chicks, you will not need a lot of food – a 50-pound (22.7 kg) bag of starter feed will cost about $15 (£13.50) last you about 6 months for three chicks. If you are raising five or six chicks, it may only last you 3 months. Once your chickens have grown up, the amount you spend on food will be determined by whether or not you allow your chickens to range freely. If you raise your chickens as free-range birds you will need to provide some supplementary feed, but your food costs will be much lower. Generally, you should budget about $10 (£9) per month for feed for a group of up to 5 Ayam Cemani chickens.

Bedding – You do not necessarily need to use bedding, especially if you allow your chickens to range freely, but it will make clean-up in your coop much easier. When you are raising your chicks in a brooder you will need to use wood shavings or recycled paper. A large bag of wood shavings costs about $15 (£13.50) and will last you several months in a brooder. In a large chicken coop, a bag this size might last you about one month. You should budget about $15 (£13.50) for bedding costs per month. If you choose to use newspaper you may not have to pay for your bedding at all.

Ayam Cemani Chickens

Chapter Three: What to Know Before You Buy

Veterinary Care – Ayam Cemani chickens are not like cats and dogs – they do not need to visit the vet regularly and they will not require any vaccination, though vaccinations are available for certain diseases. If you do need to take a chicken to the vet, however, it will probably cost you about $50 (£45). If you have to take one chicken to the vet during the year the cost will average to about $4 (£3.60) per month over the course of the year. For the sake of budgeting, set aside about $5 (£4.50) per month per chicken for emergency veterinary costs.

Repairs/Replacements – Over time you will need to make repairs to your chicken coop or to the fence, if you choose to use one to confine your Ayam Cemani chickens. You may also need to pay for replacements of feeders, waterers, brooding lamps, and other equipment. You will not have to cover these costs each and every month but, just to be prepared, budget a cost of $10 (£9) per month for unexpected repairs and replacements.

To help you get an idea how much the monthly costs for Ayam Cemani chickens are, consult the chart below. You will notice that there are cost estimates for a single chicken and for a group of 3 and a group of 5.

Chapter Three: What to Know Before You Buy

Monthly Costs for Ayam Cemani Chickens			
Cost Type	**1 Chicken**	**3 Chickens**	**5 Chickens**
Food	$10 (£9)	$10 (£9)	$10 (£9)
Bedding	$15 (£13.50)	$15 (£13.50)	$15 (£13.50)
Veterinary Care	$5 (£4.50)	$15 (£13.50)	$25 (£22.50)
Repairs/ Replacements	$10 (£9)	$10 (£9)	$10 (£9)
Total	$40 (£36)	$50 (£45)	$60 (£54)

* U.K. costs are estimated based on an exchange rate of $1/£0.90. Rates are subject to change.

Chapter Three: What to Know Before You Buy

5.) Pros and Cons of Ayam Cemani Chickens

As is true of all animals, the Ayam Cemani chicken breed has both its associated advantages and disadvantages. Before you decide whether this is the right breed for you, you should consider the pros and cons. Below you will find a list of pros and cons associated with keeping Ayam Cemani chickens:

Ayam Cemani Pros:

- Very attractive breed – unique in appearance.
- Average-sized, does not require an excessive amount of space.
- Very friendly and gentle temperament.
- Fairly easy to care for – not high-maintenance.
- Inquisitive breed, can be very interesting to keep.

Ayam Cemani Cons:

- Very expensive to obtain – could cost hundreds of dollars for a chick, thousands for an adult.
- Still a rare breed, might be difficult to obtain.
- Fairly strong-winged, could fly out of an open-top enclosure.

Chapter Four: Purchasing Ayam Cemani

Once you have made the decision that the Ayam Cemani chicken is the right breed for you, your next step is to figure out where to buy them. Because these chickens are still fairly rare in the United States and Europe, you may not be able to just go to your local poultry farmer and ask for Ayam Cemani chicks – you will have to find a specialized breeder. In this chapter you will receive tips for finding an Ayam Cemani breeder and for selecting chicks that are in good health.

Chapter Four: Purchasing Ayam Cemani Chickens

1.) Where to Buy Ayam Cemani Chickens

If you are looking for adult chickens or chicks to raise yourself, you may be able to find a local poultry farmer to supply you with the eggs or chicks. If you have your heart set on the Ayam Cemani breed, however, the task may not be that simple. These chickens are still fairly rare outside of their native country so you will probably have to find a specialized Ayam Cemani breeder. In this section you will receive tips for finding an Ayam Cemani breeder as well as some tips for ensuring that the breeder you buy from is reputable and reliable.

a.) Buying in the U.S.

In the United States, the American Poultry Association (APA) is the main organization that puts on poultry shows, organizes breeders, and provides references for poultry farmers. The Ayam Cemani is not a recognized breed with the APA, but you may be able to find an Ayam Cemani breeder through this organization. The APA also provides general information about selecting a poultry breeder and raising chicks to maturity.

If you cannot find an Ayam Cemani breeder through the APA, your next best bet is to perform an online search. When searching for poultry breeders online you need to be

Chapter Four: Purchasing Ayam Cemani Chickens

able to discern which breeders are reputable and experienced from those that are not. When performing your search, compile a list of breeders then go through each breeder's website to determine whether they are reputable or not. Contact the breeders individually to find out how long they have been breeding Ayam Cemani chickens, how much they know about the breed, and what their breeding practices are like. Any breeder that seems hesitant to answer your questions should be flagged and you should move on to the next breeder on your list.

To get you started in your search for Ayam Cemani chickens, consider some of the USA breeders below:

Greenfire Farms.
www.greenfirefarms.com/chicken/ayam-cemani

Cemani Farms.
www.cemanifarms.com/p/ayam-cemani.html

Watson Ridge Ranch.
www.watsonridgeranch.webs.com/ayamcemani.htm

Chapter Four: Purchasing Ayam Cemani Chickens

Onagadori South Feather Farm.

www.countrywhatnotgardens.com/bantamlongtails/cemani.html

Graceful Chickens.

www.gracefulchickens.com/indonesian-ayam-cemani

b.) Buying in the U.K.

Buying Ayam Cemani chickens or chicks in the U.K. is very similar to buying them in the United States. You are unlikely to find these chickens at a regular poultry farm – you will probably have to search specifically for a specialized Ayam Cemani breeder for chicks or adult chickens. You might be able to find a breeder through the British Poultry Council or the Poultry Club of Great Britain, though the Ayam Cemani is not a recognized breed by either one of these organizations.

To get you started in your search for Ayam Cemani chickens, consider some of the U.K. breeders below:

Peggy's Pekins.

www.peggyspekins.co.uk/meet-the-ayam-cemani.html

Chapter Four: Purchasing Ayam Cemani Chickens

Acorn Poultry Breeder.

http://www.chickens.allotment-garden.org/poultry-suppliers/poultry-breeder-9301.php

The Poultry Emporium.

www.thepoultryemporium.co.uk

"Stock for Sale." AyamCemani.co.uk.

www.ayamcemani.co.uk

Chapter Four: Purchasing Ayam Cemani Chickens

2.) How to Select a Healthy Ayam Cemani Chicken

If you have decided that the Ayam Cemani breed is right for you, you must then go through the process of finding and selecting a reputable breeder. Depending where the breeder is located, you might also want to pay them a visit to determine whether or not the breeder is experienced and whether the stock is healthy. The best way to ensure that your Ayam Cemani chicks grow up big and strong is to start with chicks that are healthy in the first place. In this section you will receive some tips for picking out a healthy Ayam Cemani chick.

When visiting an Ayam Cemani breeder, always check the following before buying:

- Visit the breeder's facilities and make sure that they are clean and properly maintained – a dirty enclosure increases the chances that the poultry are sick or in poor condition.

- Check the behavior and overall appearance of the breeding stock – the adult birds should appear healthy and active, not lethargic, dirty, or in poor condition.

Chapter Four: Purchasing Ayam Cemani Chickens

- Look around the holding facilities to make sure there are no signs of diarrhea – diarrhea is a common symptom of disease in chickens.

- Evaluate the chickens themselves – look for accumulated feces on the vent, signs of missing feathers or mites, discharge or pus on the eyes and/or nose, etc.

- Observe the birds to evaluate their breathing – it should not be labored and there should be no wheezing or sneezing.

- Make sure the eyes of the birds are bright and clear – Ayam Cemani chickens have dark eyes, but they should still be clear and free from discharge.

- Examine the feather coverage of the chickens – if you are buying chicks they will probably still have lots of down, but adult birds should have full coverage.

- Look over the legs and feet for signs of scaly leg (raised or encrusted scales) as well as signs of mites or injury.

- Look for the breast bone – if it is protruding very prominently it is a sign that the bird is underweight which could be an indication of poor health.

Ayam Cemani Chickens

Chapter Four: Purchasing Ayam Cemani Chickens

After you've examined the facilities and the chickens themselves, ask the breeder what kind of veterinary treatment the chickens have received. Ideally, they should be treated with some kind of de-wormer like Flubenvet. If you decide to purchase chickens from the breeder, make sure you quarantine them for three weeks before introducing them to other poultry to make sure they aren't carrying any communicable diseases.

Chapter Five: Caring for Ayam Cemani

After you have taken the time to decide whether the Ayam Cemani is the right breed for you and you have selected a reputable breeder, then comes the real work of actually raising your chickens. When it comes to raising chickens you need to decide whether you want to keep them in a coop or allow them to range freely. Even if you choose to let your chickens range freely, you should still provide them with some kind of shelter and a place to roost. In this chapter you will learn how to decide which option is best for your chickens and you will learn how to care for your Ayam Cemani chickens properly.

Chapter Five: Caring for Ayam Cemani Chickens

1.) Habitat Requirements

When you go to the grocery store to buy eggs, you have probably seen several different options. The more expensive eggs are often labeled "cage-free" or "free-range," but what does this really mean? Commercial poultry farms keep their chickens in tight, cramped spaces and they often pump their chickens full of antibiotics to make them grow fat and to speed up their growth. Free-range chickens are allowed a little more space with access to the outdoors which means that the birds are healthier (because they receive a more varied diet and they receive plenty of exercise) – this also means that the eggs are more nutritious. Free-range and cage-free eggs also taste much better than commercially produced eggs!

When it comes to raising Ayam Cemani chickens, you have several options to choose from. One option is to keep your chickens in a coop – an enclosed space that limits the range of your chickens but also protects them from the elements and from predators. Another option is to keep your chickens in a fenced-in area with access to a coop where they can sleep at night and retreat during inclement weather. If you choose this method you should go with a portable-style coop – this will enable you to move the coop every few days to renew food sources for your chickens and to fertilize the ground in different areas. If predators are not a problem in your area, you can even choose to let your

Chapter Five: Caring for Ayam Cemani Chickens

chickens range completely free with no fence at all – you should still provide a coop or some roosts/nests where your chickens can rest at night.

a.) Space Requirements for Chickens

In determining how much space to give your Ayam Cemani chickens, keep in mind that bigger is always better. The more space your chickens have to roam, the healthier and happier they will be. If you choose to keep your chickens in a coop or a fenced area, it is very important that you meet the minimum requirements for space. Most poultry farmers will recommend a minimum of 3 square feet (90 sq. cm) of ground space per chicken. This measurement refers to the amount of floor space only – it does not include nest boxes, perches, or roosting areas.

In addition to providing at least 3 square feet (90 sq. cm) of ground space per chicken, you also need to provide perches for your chickens to roost at night. Smaller to medium-sized birds like Ayam Cemani chickens need about 12 inches (30 cm) of perch space each. Keep in mind that your chickens will likely roost close together, but you still need to provide at least 12 inches (30 cm) per bird. You may also want to provide several perches at different heights so your chickens can choose the perch they like best.

Chapter Five: Caring for Ayam Cemani Chickens

Finally, you also need to provide your chickens with nest boxes – that is, if you plan to keep hens. Nest boxes should be installed below the height of your perches to make sure that your chickens don't soil them by roosting in them at night. You should provide a minimum of two nest boxes for your chickens with one nest box available for every four hens. Line the nest boxes with straw (not hay) so the hens have comfortable material to lay in. Do not use hay because it can foster mold spores which might make your chickens sick and which could affect the eggs.

b.) Types of Chicken Coops

Chicken coops come in all shapes and sizes, made from a wide variety of different materials. Some of the most popular options for chicken coop materials include plastic chicken coops, eco chicken coops, and wooden coops. There are also different shapes/style options like arks, chicken houses, poultry sheds, and portable coops. You will find a brief explanation of each option below:

Plastic Chicken Coops – This type of coop is constructed from heavy-duty plastic materials which makes them very durable. These coops are also very easy to clean and they are generally portable as well. The problem with this type of coop is that they often come with a very small run which doesn't give your chickens much room to range. These

Chapter Five: Caring for Ayam Cemani Chickens

coops also need to be used on flat ground so there are no weak points in the enclosure which might allow a predator to get inside the coop.

Eco Chicken Coops – This type of chicken coop is constructed from recycled materials which makes it an eco-friendly option. Eco chicken coops are usually bolted together and made with thick walls for optimal durability as well as easy cleaning. If you are looking for a pre-fabricated chicken coop, these are a great option because they are usually flat packed for easy shipping and transport. This type of coop doesn't usually come with a run so you have the option to build a large run around it to suit the size of your flock.

Wooden Chicken Coops – One of the most cost-efficient materials to use in making a chicken coop is wood. You can purchase pre-fabricated wooden chicken coops to assemble yourself or you can build your own. These coops come in a wide range of sizes and shapes so you can choose one that works for your flock. Keep in mind when buying a pre-fabricated wooden coop that manufacturers often have lower space recommendations – buy the coop a little larger than recommended to make sure your chickens have enough space to roam and roost.

Chapter Five: Caring for Ayam Cemani Chickens

When it comes to chicken coop designs, there are many options to choose from. Some of the most popular options include the chicken ark, the traditional chicken house, the poultry shed, and the portable coop. <u>You will find a brief explanation of each option below</u>:

Chicken Ark – A chicken ark is a small enclosure that is designed to keep your chickens completely contained while also giving them some space to move around. This type of coop is particularly popular in areas where chickens cannot be allowed to roam freely due to city ordinances or predators. Chicken arks are recommended for flocks containing fewer than 12 birds and they are typically movable so you can give your birds access to fresh grass every few days. A chicken ark typically has two enclosed ends with a fenced-in space in the middle – these coops can be triangular or rectangular in shape.

Traditional Chicken House – The traditional chicken house is based on a design created by Tarter Farm and Ranch Equipment that dates back to 1895. This design consists of a large wooden structure (similar in appearance to a barn) that includes indoor roosting space with a door that provides access to a fenced-in chicken run. You can find plans to build your own traditional chicken house below:

www.tarterusa.com/mammys-1895-chicken-house-plans-3/

Chapter Five: Caring for Ayam Cemani Chickens

Poultry Shed – The poultry shed is another very basic design that is easy to build yourself. This design is completely enclosed with several roosts, nesting boxes and feeders/waterers included as well as a closed-off area for supply storage. The shed has a sloping roof and several windows in the front to provide natural sunlight. You can provide your birds with access to the outdoors by fencing in an area around the shed and by keeping the door to the shed open.

Portable Coop – The portable coop is a great option for modified free-range chickens because it can be moved around to different areas along with your portable run. If you allow your chickens to range completely free, a portable coop provides them with a safe haven to retreat to at night. Portable coops can be built from a variety of materials and they are often built on skids or with wheels to facilitate easy movement.

The type of enclosure and coop you choose for your chickens is entirely up to you. If you have a great deal of outdoor space and no threat of predators, you can let your Ayam Cemani chickens range freely while providing them access to a coop at night. If predators are a problem, or if city ordinances prevent you from letting your chickens range freely, a modified free-range setup might be better.

Ayam Cemani Chickens

Chapter Five: Caring for Ayam Cemani Chickens

This option includes a large fenced-in area (or a portable run) along with some kind of coop. If you have very limited space for your chickens, a very small flock, or a high threat of predators, keeping your chickens entirely in a coop with a small run might be the best option.

c.) Raising Chickens in Urban Areas

When raising Ayam Cemani chickens in an urban area, you have several important factors to consider. First, you need to check with your local council to see if there are any city ordinances which might affect the number of chickens you can keep and/or the type of enclosure you need to use. Second, you need to think about your neighbors. You will need to keep your chickens and their coop clean and do what you can to minimize noise.

In addition to considering city ordinances and the preferences of your neighbors, you also have to think about how your chicken coop will fit in with the neighborhood. You do not want to use something that sticks out like a sore thumb because your neighbors will be more likely to file complaints. Make sure that your coop blends in with the style of existing buildings and take steps to make sure that you can open and shut the enclosure at specific times to minimize noise – your neighbors won't like being woken up at 5 am by your chickens.

Chapter Five: Caring for Ayam Cemani Chickens

Even if you have a lot of outdoor space, if you are living in an urban area you probably will not be able to let your chickens roam freely. Few people appreciate having chickens running around without limits and free-range chickens are at risk for predation by stray dogs and varmints in urban settings. Use strong fencing and consider installing security lights to keep your chickens safe – you should also plant shrubs around the coop to help deaden noise and to camouflage the enclosure.

When raising chickens in an urban area, it is very important that you start out small until you learn how many birds you can handle and what kind of space you have available. Start with two or three hens and make sure you have a plan to sell or use all of the eggs they will produce. Keep in mind that the eggs will not be fertile unless you have a rooster as well – keeping hens only is a good way to keep your flock size low in urban areas. If you keep a rooster you will have to deal with hatching, culling, and raising chicks. The more you plan and prepare for raising chickens in an urban area, the better.

Chapter Five: Caring for Ayam Cemani Chickens

2.) Building a Chicken Coop

Before you start making plans to build your own chicken coop, you need to decide how many chickens you want to keep and how much space you have. Determine how many chickens you want to keep and then multiply that number by 36 inches (90cm) to determine the total square footage of the floor of the coop. Keep in mind that bigger is always better when it comes to chicken coops and that you will have to add additional space for your nesting boxes if you plan to keep hens. Plan for at least two (ideally three) square feet of floor space per chicken.

Once you've determined the dimensions of your chicken coop, you can draw out the design on graph paper. You may also want to mark the ground outside where the chicken coop will be placed so you can determine which side will get the most sun. You want to make sure the open side of the coop and the run have southern exposure to take advantage of the natural warmth and sunlight – this is especially important during the winter. Don't forget to include a door in your plan as well as a mesh-lined window or two for ventilation.

After you've drawn up your plans you can take them to a lumber yard where someone will be able to help you determine how much wood you need as well as any tools

Chapter Five: Caring for Ayam Cemani Chickens

that will be required. You should plan to use 2-by-4's to frame the chicken coop, using sheets of plywood for the walls. You can use another sheet of plywood framed with 1-by-2's for the door and attach it using metal hinges. The roof of your coop can be constructed from a piece of sheet metal or plywood covered with roof shingles or roofing felt.

In addition to building your chicken coop, you should also draw up plans for the chicken run that will be attached to it (if you are going with a modified free-range plan). You can use 1-by-2's or 2-by-4's to frame the run and enclose it with sturdy 1-inch wire mesh. A run measuring 5-by-20 feet (1.5-by-6 m) should be sufficient for a small flock of 6 to 8 Ayam Cemani chickens. If you have extra room, however, more space is always better. When you construct the run, make sure to bury at least six inches of the wire in the ground to make sure that digging varmints like foxes and weasels can't get into the run. You should completely enclose the run on the sides as well as the top.

When you are finished building the coop and run, accessorize it accordingly with your nesting boxes, feeders, and waterers. You will need one nesting box for every four hens, using a minimum of two for the flock. Make sure you buy waterers and feeders, or a feed trough, large enough for all of your chickens to feed at the same time. Inside the coop, line the floor with about six inches of wood shavings

Chapter Five: Caring for Ayam Cemani Chickens

and line the nesting boxes with straw or extra wood shavings.

Don't be afraid to get creative with your DIY chicken coop. Feel free to design it so that it attaches to an existing structure like a shed or your house. You can also create a portable chicken coop by building it on skids and attaching wheels. When it comes to the shape of your coop you can stick with a standard rectangular shape or choose a triangular shape for simplicity sake. Just remember that you need to provide a minimum of 3 square feet (90 sq. cm) of floor space per bird and you will be fine.

Chapter Five: Caring for Ayam Cemani Chickens

3.) Maintaining Your Chicken Coop

When it comes to maintaining your chicken coop, your main task will be to replace the bedding at least once a fortnight, as needed. Chickens can be quite dirty so replacing the bedding will not only help to keep the coop clean but it will also minimize the risk for mold, parasites, and disease transmission. Not only do you need to refresh the bedding on the floor of your coop, but also replace the bedding in the nesting boxes and below the perches/roosts. Ideally, you should choose a coop design that has removable nest boxes to facilitate easy cleaning.

In addition to replacing the bedding in your coop, you should also wash the coop out once in a while. Use a specialized poultry detergent to help make sure the coop is properly cleaned and to avoid exposing your chickens to any harmful chemicals. A poultry detergent like Poultry Shield will also help to reduce the risk for parasites like red mites. Cleaning out your coop will be especially important during the warm summer months when temperatures rise. Keeping your coop properly ventilated will also help to maintain sanitary conditions all year round.

Chapter Five: Caring for Ayam Cemani Chickens

4.) Keeping Your Chickens Warm

Like most chickens, the Ayam Cemani chickens can withstand fairly low temperatures as long as they have indoor space. If you live in an area that gets very cold in the winter you will need to take certain precautions to keep your chickens warm. You will also need to make sure that their enclosure is draft-free and well ventilated to prevent the spread of disease in closed quarters.

To keep your chickens warm in the winter, you may need to keep them in the coop for most of the time. If you do, however, you will need to put some ventilation measures in place – ventilation windows are a great option because they help to release fouled air while bringing in fresh air. Just be sure to point the windows away from the wind and close them if wind starts blowing into the coop.

Chicken droppings release ammonia as they decompose which produces heat and which affects air quality – having ventilation windows in place will help to filter the ammonia-laden air out, bringing fresh air in. Without proper ventilation, your chickens could breathe in too much ammonia-laden air which will damage their lungs as well as their eyes. As a whole, chickens are very susceptible to respiratory problems so proper ventilation is incredibly important. The cleaner and more ventilated your coop is,

Chapter Five: Caring for Ayam Cemani Chickens

the less stressed your chickens will be which will also play a role in keeping them healthy.

Chapter Six: Feeding Ayam Cemani Chickens

After going through the process of designing your chicken coop you may be eager to get started raising your Ayam Cemani chickens. Before you begin, however, there is still something very important you have to learn – how to feed your chickens. The nutritional needs of chickens are different from those of other birds so you need to learn exactly what your chickens will need to eat before you go out and buy them. In this chapter you will learn the basics about nutritional needs for Ayam Cemani chickens and receive tips for feeding your chickens a healthy diet.

Chapter Six: Feeding Ayam Cemani Chickens

1.) *Nutritional Needs of Chickens*

If you have ever kept chickens before you may already have noticed that they never stop moving. Chickens are always on the move which means that they have very high needs for nutrients and for energy. In order to ensure that your Ayam Cemani chickens remain healthy and active you need to familiarize yourself with their basic nutritional needs so you can provide a diet that meets those needs. The nutritional needs for Ayam Cemani chickens are not significantly different from the needs of other domesticated breeds. The information provided in this chapter is based on nutrient requirement figures published by the National Research Council found in the Merck Veterinary Manual.

The nutrient requirements for Ayam Cemani chickens change as these birds grow from hatchlings into adult birds. The energy needs of chickens vary according to their environment and their level of physical activity. Chickens are very good about regulating their diet – they stop eating when they have consumed enough nutrients and they increase their food intake if their needs have not yet been met. This being the case, you do not need to worry about overfeeding your chickens – your main concern should be to provide them with enough food that they can eat as much as they want/need to.

Chapter Six: Feeding Ayam Cemani Chickens

Like all animals, Ayam Cemani chickens require a balance of protein, carbohydrate, and fat in their diets. They also require certain vitamins and minerals. The protein needs of chickens are fairly high during the development period and they rise again during laying. Chickens can produce certain amino acids (the building blocks of proteins) themselves but require certain other amino acids in their diet – these include arginine, lysine, methionine, and threonine to name a few. Ayam Cemani chickens also require certain vitamins and minerals in their diet, including calcium and phosphorus. Most of the nutritional needs of chickens can be met by feeding them a commercial diet.

To give you an idea what kind of nutrients a chicken needs to be healthy, consult the chart below:

Nutritional Needs for Chickens			
Chicks			
Age (in weeks)	**0 to 6**	**6 to 12**	**12 to 18**
Body Weight	**450g**	**980g**	**1,375g**
Protein	18%	16%	15%
Arginine	0.94%	0.78%	0.62%
Lysine	0.80%	0.56%	0.42%
Methionine	0.28%	0.23%	0.19%
Threonine	0.64%	0.53%	0.35%
Tryptophan	0.16%	0.13%	0.10%
Calcium	0.90%	0.80%	0.80%
Phosphorus	040%	0.35%	0.30%

Chapter Six: Feeding Ayam Cemani Chickens

Laying Hens			
Protein	18.8%		
Arginine	0.88%		
Lysine	0.86%		
Methionine	0.38%		
Threonine	0.59%		
Tryptophan	0.20%		
Calcium	4.12%		
Phosphorus	0.31%		

In addition to the essential nutrients that have already been mentioned, Ayam Cemani chickens also need a great deal of water. Make sure that your chickens have unlimited access to fresh, clean water at all times, especially during their early growth and development. In most cases, a chicken will drink two to three times as much water by weight as they consume in feed. Water consumption will increase during the warmer months and decrease a little bit in the winter. If it gets very cold in the winter, remember to check their water regularly as it may become frozen.

Chapter Six: Feeding Ayam Cemani Chickens

2.) Types of Chicken Feed

If you pay a visit to your local farm supply store you will notice that the shelves are stocked with many different kinds of chicken feed. Not only are there half a dozen or so different standard formulas, but each brand is a little bit different from the others. In this section you will learn about the different types of chicken feed and when you should be using them for your Ayam Cemani chickens.

Starter Feed

When you bring your Ayam Cemani chicks home or if you hatch them yourself you will need to put them on a starter feed. These formulas typically contain between 18% and 20% protein as well as a specific combination of nutrients that chicks need to facilitate healthy growth and development. If you are raising your Ayam Cemani chickens for meat, you may even consider using a starter diet formulated for meat chickens – these formulas are a little higher in protein, around 22%, to help maximize growth of your chickens.

Grower Feed

Once your chicks reach about 6 weeks of age you should switch them to a grower feed to help them sustain a steady growth rate until they reach maturity. These formulas typically contain 15% to 16% protein and you can feed your

Chapter Six: Feeding Ayam Cemani Chickens

chicks this formula until they are about 14 weeks old. At this point you should switch them to a developer feed – these formulas contain 14% to 15% protein and they are ideal if you are preparing your chickens for egg production. It is not necessary to switch your chicks to a developer feed but once they reach 20 weeks you will need to swap over to an adult feed formula.

Layer/Breeder Feed

For Ayam Cemani hens, you will have to choose between a layer or a breeder formula. Layer formulas contain about 16% protein and they have extra calcium to ensure that the chickens lay eggs with strong shells. This type of formula is recommended if you are keeping chickens to produce eggs for food. You should start your chicks on a layer formula once they reach about 20 weeks of age. If you intend to hatch your Ayam Cemani eggs, your hens should be kept on a breeder formula. These formulas have a little bit more protein than layer formulas and they are supplemented with extra vitamins to ensure proper development and hatching of the eggs.

Medicated Feed

Another type of feed you might consider using is the medicated feed. Most commercially produced starter feeds are medicated to help prevent some of the most common diseases known to affect flocks of chickens. It is less

Chapter Six: Feeding Ayam Cemani Chickens

common for grower feeds and layer feeds to be medicated, though you can still find them. Always read the label carefully before using a medicated feed, especially if you are feeding it to laying hens. The label will give you instructions for when to withdraw the medicated feed if there is a risk of the medication being passed to the eggs. As long as you follow the instructions on the package, however, you do not have to worry about the eggs being safe to eat.

Chapter Six: Feeding Ayam Cemani Chickens

3.) Other Types of Food

In addition to commercial chicken feeds, you should also be offering your Ayam Cemani chickens other types of food including scratch grains, green vegetables, fruit, and table scraps. In this section you will receive recommendations for feeding each of these types of food.

Scratch Grains

Even if you have never kept chickens before you are probably familiar with their scratching behavior. Chickens scratch the ground with their feet to break up the dirt, looking for seeds, insects, and grit to eat. Offering your chickens scratch grains can help to encourage this type of behavior which will also help to make sure that your chickens get enough exercise. Examples of scratch grains include cracked, rolled or whole corn, barley, wheat, and oats. These grains are fairly low in protein but high in dietary fiber.

Most poultry farmers recommend using limited amounts of scratch grains – if you use too much it could affect your chickens' appetites for more nutritious foods. You should only offer your chickens scratch grains in the afternoon after they have eaten their regular feed. Only offer your chickens a few handfuls of scratch grains (depending how many chickens you have), limiting your feeding to an

Chapter Six: Feeding Ayam Cemani Chickens

amount they can eat within 15 or 20 minutes. You can offer a little more during the winter to provide your chickens with extra calories to stay warm.

Grit

Chickens do not have teeth so they eat grit to help grind down food in their gizzards. It is especially important to feed your chickens grit if you offer them scratch grains because these foods are harder to digest. Free-range chickens can often find enough grit on the ground in the form of small rocks and pebbles but it never hurts to offer some anyway. In addition to grit you might also want to offer some crushed oyster shell because it is high in calcium. Most commercial layer feeds contain an adequate amount of calcium for laying hens but it never hurts to provide a secondary source.

Green Vegetables

In order to keep your Ayam Cemani chickens in good health, you must feed them a varied diet that consists of more than just commercial feed pellets – fresh greens should also be part of the diet. Free-range chickens will be able to feed on weeds and other plants but you should still offer some vegetable greens on a daily basis. Some of the best vegetables to offer your chickens include cauliflower leaves, cabbage, spinach, dandelion greens, and grass

Chapter Six: Feeding Ayam Cemani Chickens

clippings. You can also feed your chickens potatoes and potato peelings as long as you boil them first.

Fruit

In addition to green vegetables you can also give your chickens small amounts of fruit as an occasional treat. Grapes and strawberries are a favorite among most chickens – just be careful not to feed too much or your chickens could develop diarrhea.

Table Scraps

The diet of chickens that are kept primarily in coops can be supplemented with table scraps for variety. Free-range chickens, on the other hand, can usually find enough food to supplement their commercial feed diets. If you do choose to offer your chickens table scraps, make sure not to feed too many carbohydrates and limit scraps to 20% or less of the birds' diet. The best way to feed table scraps is to stir them up with some layers mash and add a little bit of water to create a crumble. Most chickens will eat almost anything – everything from coffee grounds to stale toast. You should be careful to never feed your chickens rhubarb leaves or avocado, however, because they are toxic to chickens.

Chapter Six: Feeding Ayam Cemani Chickens

4.) Ayam Cemani Feeding Recommendations

For the first two days after Ayam Cemani chicks are born, they continue to absorb what is left of the yolk from the egg. After two days, however, you will need to make food and water freely available. Offer your newly hatched chicks a finely ground starter feed free-choice until they are 14 to 18 weeks old. At that point you can switch to a layer or breeder feed. Adult chickens are very good at regulating their intake of food so you generally do not need to worry about overfeeding. Just make sure that you have enough feeder space available to accommodate your entire flock at one time as well as plenty of waterers.

In addition to learning how to properly feed your Ayam Cemani chickens, you should also take the time to learn about some common feeding mistakes so you can avoid making them yourself. One of the most common feeding mistakes poultry farmers make is giving their chickens too much mixed corn. Mixed corn usually contains about 80% to 90% wheat with just 10% to 20% cracked maize. Wheat is relatively low in protein (around 10%) which makes it a poor feeding choice for laying and breeding hens. Maize is also fairly high in fat – this may be okay for feeder chickens but it is not ideal for laying hens.

Chapter Seven: Breeding Ayam Cemani

Breeding Ayam Cemani chickens is not a task that should be undertaken lightly. Because these birds sell for such a high price it can be tempting to breed them simply for profit. It is important to realize, however, that breeding poultry can be a fairly complex process. Not only do you need to learn how to incubate and hatch the eggs, but you also have to learn about culling – this is a necessary evil for any poultry breeder. In this chapter you will learn the basics about breeding requirements for Ayam Cemani chickens and receive tips for incubating, hatching, and culling the chicks.

Chapter Seven: Breeding Ayam Cemani Chickens

1.) Basic Breeding Info

Before you make the decision to breed your Ayam Cemani chickens you should consider your reasons for doing so. Do you simply want to make an extra buck or are you legitimately interested in growing a flock of healthy and vibrant Ayam Cemani chickens? Unless you are willing to do the work to properly breed your chickens and care for the chicks, you should not breed them. Your best bet is to keep a small flock of Ayam Cemani hens and to use them for egg production purposes only.

If you do decide that breeding your chickens is the right choice you need to take the time to learn the basics about chicken breeding. Your first step is to choose the two chickens you want to breed – one male and one female. If you are breeding your chickens simply to build your own flock it may not be important which birds you choose as long as they are both healthy. If you are breeding your Ayam Cemani chickens for show, however, you need to select the ideal specimens of both sexes to use in breeding so that the chicks will be good examples of the Ayam Cemani breed standard.

Ayam Cemani chickens do not lay as many eggs each year as other domesticated breeds so the breeding process could be a little lengthier than you might have otherwise

Chapter Seven: Breeding Ayam Cemani Chickens

imagined. The average Ayam Cemani hen lays about 60 eggs per year and, if you plan to hatch the eggs, they will all have to be artificially incubated. You will learn more about incubating your eggs later in this chapter.

Chapter Seven: Breeding Ayam Cemani Chickens

2.) The Breeding Process

The breeding process for Ayam Cemani chickens is not significantly different from that of other domesticated breeds. If you place a mature male chicken (a cock) in the same area as a group of mature females (hens), breeding will occur with little input from you. From very shortly after the two sexes have been introduced you should start to notice certain courting behavior. The male will initiate courting behavior by dropping one of his wings and dancing in a circle. In response, the hen will crouch down, dipping both her head and her body, as an indication that she is receptive to mating.

Once the hen indicates her receptiveness to mating, the male will mount her, grabbing her by the neck, comb, or back to hold himself on. You will then witness the male treading on the hen's back followed by a dipping of his tail to the side of the hen's tail as he spreads his tail feathers. When the cloacae of both birds come into contact with each other, the male releases his ejaculate directly into the female's vagina where fertilization might occur.

Ayam Cemani hens will lay eggs whether or not they have been fertilized. It is only the fertilized eggs, however, that can be hatched into chicks. It takes about 25 hours for an egg to make its way through the hen's system and

Chapter Seven: Breeding Ayam Cemani Chickens

fertilization can only take place during the first 15 minutes or so after the egg has been released. This means that fertilization may not occur with every mating and it may take several mating attempts for a male Ayam Cemani chicken to fertilize an egg. In most cases, however, Ayam Cemani hens will remain fertile for about two weeks after a mating. If you want to make sure that your hen is only fertilized by a particular rooster you should keep her separated from other roosters for at least 3 weeks.

Chapter Seven: Breeding Ayam Cemani Chickens

3.) Hatching Ayam Cemani Eggs

Whether you breed your own Ayam Cemani chickens or receive your eggs from another breeder, you have to be very careful about how you go about hatching them. Ayam Cemani hens are not sitters so they will not incubate the eggs themselves – you have to incubate them artificially. If you order your eggs by mail you should let them settle for about 24 hours or so before placing them in the incubator. This rest period is necessary to allow the air inside the eggs to normalize before incubation.

In addition to giving your Ayam Cemani eggs a little bit of time to normalize, you can store them for up to 14 days before putting them in the incubator. After 14 days the rate of hatchability declines significantly. To store your eggs, keep them in an egg carton or another egg storage container at a temperature of 55° to 60°F (13° to 15.5° C) and a humidity between 70% and 75%. Keep the eggs stored pointy side down and slant or turn them daily during storage. To make this easy you can place a piece of 2-by-4-inch wood under one end of the carton and change it to the other side once a day.

Before you start incubating your eggs, you should have the incubator running for at least 24 hours – this will ensure that it is the proper temperature when you add your eggs.

Chapter Seven: Breeding Ayam Cemani Chickens

The ideal temperature for an Ayam Cemani incubator is 37.5°C (99.5°F). The humidity should be kept at 50% for the first 18 days and then increased to 55% to 60% for the last three days that the eggs spend in the incubator.

It is very important to keep a steady temperature in your incubator so you should use a thermometer to monitor the temperature. It will be easier to maintain a steady temperature in a larger incubator and slight changes in temperature will be less devastating. You should make adjustments to the temperature as needed judging by the results of each hatching. If the eggs hatch too early, lower the temperature a little for the next batch. If they hatch too late, raise the temperature a little.

To maintain the proper humidity in your incubator you'll have to use a hygrometer to get an accurate reading. Again, maintaining a stable humidity level will be easier in a large incubator. As long as you keep the humidity within 10% or 15% of the ideal range, however, your eggs should be just fine. Temperature is much more critical than humidity when it comes to hatching your Ayam Cemani eggs. As long as you maintain the right conditions your eggs should hatch in about 21 days. Make sure to turn the eggs three times a day during the first 18 days and then leave them alone for the last 3 days before hatching. After the 18th day, keep the incubator closed unless you need to add water to maintain the humidity.

Chapter Seven: Breeding Ayam Cemani Chickens

4.) How to Deal with Culling

If you decide to breed your Ayam Cemani chickens you may have to sell some of the eggs or chicks at some point. Although this breed of chicken doesn't produce as many eggs as some domesticated breeds, you probably do not have the capacity to add 50 or more birds to your flock each year. Selling some of the eggs or young chickens is a much nicer alternative to culling.

Keep in mind that it takes 5 to 6 months for Ayam Cemani chicks to mature enough to start laying their own eggs. Ayam Cemani hens do not lay daily eggs like some breeds, but you should learn how often your hens lay and use that information to determine how many birds you need to keep to meet your family's need for eggs. As you hens get older they will produce less and less eggs.

If you decide to cull your older hens rather than keep them as pets, you have two options – you can do it at your hen's first molt or wait until the second molt or later. Hens experience their first molt around 18 months of age. During this process they lose most of their feathers and grow new ones. This process takes a lot of extra protein and calories which disrupts egg production. It can take 3 months or more for the hen to begin laying again at a normal rate. Some hens stop laying completely during their molt whilst

Chapter Seven: Breeding Ayam Cemani Chickens

some hens continue to lay, though often at a reduced rate. If you don't want to deal with increased food costs and decreased production, start raising replacement hens about 6 months early and cull the older hens when they start to molt.

Your second option is to cull the hens after their second molt or later. If you do wait for your hens to complete their first molt, the eggs they lay after will be a little bit larger. It is also important to note, however, that those eggs will come fewer and farther between. Your Ayam Cemani hen will eventually get to an age when they will stop producing eggs altogether at which point you may decide to cull the bird or keep it as a pet. If you decide to keep your hens as pets they can live for 8 to 10 years.

The most common method used for culling chickens is the neck dislocation method. When performed correctly, this method renders the chicken unconscious immediately – this minimizes its suffering during the process. Another obvious way is to take the chicken to a vet where they will be able to put the bird to sleep for you.

Below you will find a step-by-step guide for culling chickens:

Chapter Seven: Breeding Ayam Cemani Chickens

1. Catch the chicken as calmly as possible – try to do it in the evening when the bird is already roosting and in its calmest state.

2. Use your non-dominant hand to hold down the chicken's legs, gripping them just above the feet.

3. Place the chicken's chest on top of your thigh to support its weight – this will result in holding the bird upside down.

4. Hold the bird's neck between your thumb and forefinger, positioning the thumb under the bird's beak and tilting its head back slightly.

5. In a firm, quick motion pull the neck sharply downward while bringing the head back at the same time – press your knuckles into the vertebrae so the neck is stretched while the head is bent all at the same time.

6. Continue to hold the bird for a few seconds until the flapping and kicking subsides – this can continue for several seconds even after the dislocation.

Chapter Seven: Breeding Ayam Cemani Chickens

If you wish to consume the meat, all that is left is to pluck and dress the bird. You can learn to do this yourself or you can take your culled chickens to a butcher to have it done for you. You can then use the meat as you like.

Chapter Eight: Keeping Ayam Cemani Chickens Healthy

In addition to providing your Ayam Cemani chickens with a clean coop and a healthy diet, you also have to take certain precautions to maintain the health and wellness of your birds. Like all animals, chickens are prone to developing certain diseases and Ayam Cemani chickens are no exception. In this chapter you will learn the basics about the most common diseases affecting this breed so you can identify them and get started with as soon as possible – that is the key to a fast recovery.

Chapter Eight: Keeping Ayam Cemani Chickens Healthy

1.) Common Health Problems

You already know that keeping your Ayam Cemani chickens on a healthy diet is the key to their wellbeing. Even if you are careful about making sure that your chickens' nutritional needs are met, however, they could still get sick. The fact of the matter is that chickens can carry and transmit diseases just as easily as any other animal. Your best course of action is to familiarize yourself with the diseases most commonly seen in chickens and to learn how to diagnose and treat them. You will find all of this information and more in this chapter.

Some of the most common diseases affecting domesticated chickens like the Ayam Cemani include:

- Avian Influenza
- Avian Pox
- Botulism
- Coccidiosis
- Egg Binding
- Infectious Bronchitis
- Infectious Coryza
- Marek's Disease
- Newcastle Disease

Chapter Eight: Keeping Ayam Cemani Chickens Healthy

Avian Influenza

Also known as avian flu, this disease is known to occur in nearly all species of birds, not just chickens. Avian influenza comes in two forms – mild and highly pathogenic. The mild form of the disease produces symptoms like lethargy, loss of appetite, diarrhea, difficulty breathing, drop in egg production, and low mortality. The highly pathogenic form of the disease causes facial swelling, dehydration, and severe respiratory distress.

The virus responsible for avian influenza can survive at moderate temperatures for long periods of time and it can live indefinitely when frozen. This means that the disease can easily be spread, even from infected carcasses or manure as well as contaminated clothing, equipment, and insects or rodents. Unfortunately, there is no treatment available for this disease and the highly pathogenic form is often fatal.

When avian influenza occurs in the mild form, it can be controlled with proper nutrition and sanitation. Treatment with broad spectrum antibiotics may help to reduce secondary infection. Even after a flock recovers it may continue to spread the virus and vaccines are only available with a special permit. Quarantine is the recommended treatment for this disease.

Chapter Eight: Keeping Ayam Cemani Chickens Healthy

Avian Pox

This disease is also called "chicken pox," though it is very different from the disease of the same name that is known to affect humans. Avian pox is known to affect most types of poultry and chickens of all ages are susceptible. There are two different forms of the disease. One is characterized by dry, wart-like lesions growing on unfeathered areas. The second is characterized by wet, canker-like lesions growing inside the mouth and throat.

The dry form of avian pox typically manifests on the legs, head, and other unfeathered areas. The lesions usually heal within two weeks, though healing may be delayed if the scab is removed before completely healed. The wet form of avian pox often leads to secondary respiratory infections which can become very serious. It is possible for a chicken to have both forms of avian pox at the same time.

Avian pox is usually transmitted by direct contact between infected and uninfected birds, though it can also be transmitted by mosquitos. The avian pox virus can enter the bird's bloodstream through the skin, eye, or respiratory tract and it tends to spread fairly slowly. Unfortunately there is no treatment for avian pox, though vaccination is possible to help stop an outbreak. The best methods for preventing this disease involve mosquito control, vaccination, and proper sanitation measures.

Chapter Eight: Keeping Ayam Cemani Chickens Healthy

Botulism

This disease is known to affect all species of fowl as well as humans and other animals. Botulism is a type of poisoning that is caused by consumption of spoiled food containing a neurotoxin that is produced by the bacteria Clostridium botulinum. The most common symptoms of botulism include paralysis and loose feathers. Paralysis can set in quickly, within hours after eating the poisoned food, and it can affect the legs, wings, and neck. In many cases, the feathers on the neck also become very loose.

If the bird consumes a lethal dose of the neurotoxin it will become paralyzed and die within 12 to 24 hours. Death is usually a result of paralysis of the respiratory muscles. If the dose ingested is not lethal it will typically cause the bird to become dull and lethargic. Botulism cannot be spread from one bird to another but it can be harbored by fly larvae that has fed on decaying carcasses or other organic matter.

Unfortunately, botulism is a very deadly disease that can kill an entire flock within a matter of hours or days. The only treatment is to remove the spoiled feed and to flush the entire flock with Epsom salt in water. The dosage rate should be 1 pound of salt per 1,000 hens in water or in a wet mash. Adding potassium permanganate to the drinking water may also be beneficial.

Chapter Eight: Keeping Ayam Cemani Chickens Healthy

Coccidiosis

The disease known as coccidiosis is caused by a parasitic organism that can be found on the ground or in contaminated feces. Once ingested, the parasite attaches itself to the lining of the bird's intestinal tract where it multiplies and begins feeding. This causes the intestines to bleed which may result in bloody stools or bleeding from the cloaca. Infected birds will spread the parasite in its feces starting several days before symptoms begin to present.

It usually takes about three days for infected chickens to start displaying symptoms of coccidiosis. In most cases, infected birds will start to droop and they will stop feeding and huddle together. By the fourth day you will begin to notice blood in their droppings – bleeding may increase until the bird dies from excessive blood loss. It is possible for an infected bird to recover after which point it will be immune to recurrent infections.

There are several treatment options available for coccidiosis in chickens, though drug-resistant strains are becoming a major problem. The best ways to prevent this disease include using medicated feed, maintaining proper sanitation, keeping litter dry and clean, and avoiding overcrowding in the coop.

Ayam Cemani Chickens

Chapter Eight: Keeping Ayam Cemani Chickens Healthy

Egg Binding

Egg binding is a condition that can affect any female bird. This condition occurs when the egg becomes stuck in the oviduct and the bird becomes unable to pass it. Fertilized eggs usually take about 20 hours for the shell to form and, once formed, it takes about an hour for the egg to move from the uterus and out of the bird's body. If the shell of the egg isn't hard enough or if the bird is dehydrated, it could have trouble passing the egg and it might become stuck.

Some of the most common symptoms of egg binding include loss of appetite, abdominal straining, watery diarrhea, pale face/comb, and hard abdomen. The hen might also make frequent trips into and out of the nest box or exhibit abnormal movement. Egg binding can be tricky to diagnose in Ayam Cemani chickens since they do not lay as frequently as other breeds. Just because a hen hasn't laid an egg for a few days doesn't mean it is egg-bound.

Egg binding is a very serious condition that can become fatal if not treated. This condition can be caused by a lack of calcium in the bird's diet or improper nutrition in general. It may also be secondary to another illness or the result of a sedentary lifestyle. Once you have diagnosed your chicken with egg binding you need to remove the egg as quickly and safely as possible – you may need to call a vet.

Chapter Eight: Keeping Ayam Cemani Chickens Healthy

Infectious Bronchitis

Also sometimes referred to as IB, infectious bronchitis is a disease that only affects chickens. This disease is caused by a virus and it is incredibly contagious. The virus can be spread through infected carcasses, unsanitary conditions, contaminated feed, even through the air. The virus can also be transmitted from a hen to her eggs, though infected eggs typically do not hatch.

The severity of this disease may vary depending on the age and immune status of the flock – it may also be influenced by environmental conditions. Infected birds usually decrease their food and water consumption and they may start making chirping sounds. Chickens with IB often produce a watery discharge from the eyes and nose along with labored breathing or gasping. In laying hens, egg production may drop significantly but will usually recover within 6 weeks, though at a lower rate.

There is no specific treatment for this disease, though antibiotics may be helpful in combating secondary infections. When brooding-age chicks are infected, raising the temperature by 5 degrees may help to reduce symptoms. There is a vaccine available for this disease and proper sanitation can help with prevention as well.

Chapter Eight: Keeping Ayam Cemani Chickens Healthy

Infectious Coryza

Also sometimes called roup, infectious coryza is a disease known to infect chickens, pheasants, and guinea fowl. This disease can be transmitted through bird-to-bird contact, even from chickens that have recovered from the disease (they still remain carriers). It is common for new additions to a flock to introduce the disease because, in many cases, infected birds don't show symptoms. Contaminated feed and water are other common modes of dispersal.

Some of the most common symptoms associated with infectious coryza include swelling of the face, foul-smelling discharge from the eyes and nose, and labored breathing. Diarrhea is another common symptom and young birds often display stunted growth. This disease is not typically fatal, but it can reduce egg production rates. The disease can last for as little as a few days to as long as two or three months.

The best treatment option for infectious coryza is antibiotics like sulfadimethoxine. Alternative treatments include erythromycin, tetracycline, and sulfamethazine. The best way to avoid this disease is to maintain proper sanitation and to avoid mixing flocks. Chickens can be vaccinated for this disease starting at 5 weeks.

Chapter Eight: Keeping Ayam Cemani Chickens Healthy

Marek's Disease

Also known as acute leucosis, Marek's disease is a type of non-respiratory viral disease. This disease is known to affect chickens between the age of 12 and 25 weeks, though older birds can be infected. Marek's disease is a form of cancer that results in the formation of tumors in the nerves which cause lameness or paralysis. Tumors can also form in the bird's eyes and internal organs – symptoms vary depending on the location of the tumors.

When tumors develop in the eyes, infected birds may exhibit irregularly shaped pupils or developed blindness. When tumors form in the internal organs, symptoms may include loss of coordination, paleness, weakness, labored breathing, and enlarged feather follicles. At the end stages of the disease, infected birds are often emaciated in appearance with pale, scaly combs.

Marek's disease can be transmitted through the air and it can live in feces, dust, and saliva. Infected birds carry the virus in their blood and can infect susceptible birds for the rest of their lives. There is no treatment for this disease, though chicks can be vaccinated to prevent tumor formation. The vaccine does not, however, prevent the virus from infecting the bird.

Chapter Eight: Keeping Ayam Cemani Chickens Healthy

Newcastle Disease

Also known as pneumoencephalitis, Newcastle disease is highly contagious and known to infect all types of birds as well as humans and other animals. Outside of avian species, the disease usually only causes mild conjunctivitis. In birds, however, Newcastle disease presents in three different forms - the mildly pathogenic form (lentogenic), moderately pathogenic form (mesogenic), and the highly pathogenic form (velogenic).

The most common symptoms of Newcastle disease include sudden onset of hoarse chirping in chicks, watery nasal discharge, facial swelling, labored breathing, paralysis and trembling. Depending on the type, mortality ranges from 10% to about 80%. In laying hens, this disease may also lead to a drop in egg production along with decreased consumption of food and water in most birds.

Unfortunately, Newcastle disease is easily transmitted short distances by air or through contaminated feed, shoes, equipment and wild animals. There is no treatment for this condition, though antibiotics can be administered for 3 to 5 days to help prevent secondary infections. There is a vaccine available for this disease and proper sanitation will help with prevention.

Chapter Eight: Keeping Ayam Cemani Chickens Healthy

2.) Preventing Illness

The key to keeping your Ayam Cemani chickens healthy is two-fold. First, you have to feed your chickens a healthy diet that meets their nutritional needs. If your chickens do not get the nutrients they need, they will fail to thrive and they may even succumb to nutritional deficiencies. Additionally, chickens that aren't properly fed may become stressed and, as a result, may have a weakened immune system. This puts them at an increased risk for contracting various diseases.

The second component involved in keeping your chickens healthy is to maintain proper sanitation. Chickens can be a little messy so you need to follow a routine to keep your coop clean. Using wood shavings or other bedding on the floor of the coop will help to collect moisture and feces – it is also easy to just sweep away the dirty bedding and replace it as needed. You should also clean and sanitize your feeders and waterers on a regular basis, especially if they become contaminated with feces.

In addition to maintaining a healthy diet and proper sanitation you should also be careful about adding new birds to your flock. It is very easy for one chicken to pass a disease on to another so you should quarantine all new birds for at least two weeks before adding them to the flock.

Chapter Eight: Keeping Ayam Cemani Chickens Healthy

This may mean keeping an extra chicken coop around. You can also use this coop to quarantine chickens that are sick to prevent them from infecting the rest of your flock.

Chapter Nine: Showing Ayam Cemani

The idea of a poultry show may sound strange to you if you have never raised chickens before, but it is actually quite common.

Chapter Nine: Showing Ayam Cemani

The challenge of showing your Ayam Cemani chickens can be very exhilarating and it is also a great opportunity to network with other poultry farmers. In this chapter you will learn the basics about showing Ayam Cemani chickens including information about the breed standard and tips for preparing for and then attending a poultry show.

1.) Breed Standard

Before you can show your Ayam Amani chicken you need to make sure that he is a good representation of the breed standard. The breed standard is a set of characteristics to which animals of a certain species or breed are compared during show. A breed standard represents the ideal specimen of the breed so you want your Ayam Cemani chicken to be as close to that standard as possible to improve your chances of winning in the show. Even if your chicken doesn't stand up to the breed standard you can still show him to gain experience.

Below you will find an overview of the breed standard for Ayam Cemani chickens divided for males and females.

Male Ayam Cemani Breed Standard:

Chapter Nine: Showing Ayam Cemani

Head and Beak – The head should be round and long, black in color, and thick. Length should be about 6.35 cm, height about 3.6 cm. The beak is black in color, length 1.8cm and width 1.2 cm. The tongue, landslide, and throat are black.

Comb and Wattles – The comb should be black and single with 3, 5 or 7 points. Wattles are large, coupled, smooth, and black in color.

Eyes – The eyes should be round and black.

Neck and Body – The neck should be medium and black, the body slanting to the back.

Chest, Back and Stomach – The chest should be wide and large, measuring 12.5cm in length and 34.1cm in diameter. The stomach is thin. The back is thin.

Feathers – The feathers are solid black over all.

Foot and Toes – The feet have long claws measuring 10cm. The toes are thick.

Wings – The wings are sturdy and strong, hard of feather and slanting.

Chapter Nine: Showing Ayam Cemani

Skin and Meat – Both the skin and meat are black.

Weight – The ideal weight is 2 to 3.5 kg (4.4 to 7.7 lbs.)

Female Ayam Cemani Breed Standard:

Head and Beak – The head should be long and flat, black in color, and thick. Length should be about 5.7 cm, height about 3.3 cm. The beak is black in color, length 1.8cm and width 0.9 cm. The tongue, landslide, and throat are black.

Comb and Wattles – The comb should be black and single with 3, 5 or 7 points. Wattles are small, coupled, smooth, and black in color.

Eyes – The eyes should be round and black.

Neck and Body – The neck should be medium and black, and thick of feather.

Chest, Back and Stomach – The chest should be wide, the stomach a width of 4 fingers and soft. The back is flat, wide, and strong.

Chapter Nine: Showing Ayam Cemani

Feathers – The feathers are solid black over all.

Foot and Toes – The feet have long claws measuring 6 cm. The toes are thick.

Wings – The wings are closed and strong, hard of feather, with a flat and slanting wingspan.

Skin and Meat – Both the skin and meat are black.

Weight – The ideal weight is 1.5 to 2.5 kg (3.3 to 5.5 lbs.)

Chapter Nine: Showing Ayam Cemani

2.) What to Know Before Showing

After you've determined that your Ayam Cemani chicken is fit for show, you then have to learn the ropes of showing poultry. Each poultry show has its own unique set of rules and restrictions, so be sure to read up on the specifics of each show before you enter. It is also a good idea to attend a few poultry shows before you enter your first one so you can get a feel for things. You can also speak to some of the exhibitioners to get tips for preparing your chickens and yourself for your first show.

One of your first tasks in preparing for a poultry show should be to pen train your chickens. At poultry shows, the birds are kept in small show cages in rows that are typically stacked two units high. This type of environment can be stressful for a bird that isn't used to it and a stressed-out bird will not look its best. Try to get your chickens used to being kept in a pen before the show so that they remain calm when show time comes around. Below you will find some tips for pen training your chickens:

- In the weeks leading up to the show, put your chickens in the pen once or twice a day.

- Get the chickens used to being handled and used to being put into and taken out of the pen.

Chapter Nine: Showing Ayam Cemani

- Place the radio or talk to your chickens while they are in the pen to get them used to ambient noise and to help keep them calm.

- Start by leaving your chicken in the pen for just a few minutes at a time and slowly work your way up to a period of several hours.

In addition to pen training your Ayam Cemani chickens you may also have to physically prepare them for the show. You should plan to wash your chickens a few days before the show in order to give their plumage time to dry naturally. You will need to scrub the bird's feet, toes, and nails to remove dirt. Once your chicken is clean you may then want to keep it in a pen with plenty of clean bedding to keep it from getting dirty again before the show.

Another aspect of poultry show preparation that shouldn't be overlooked is nutrition. A few months before the date of the show you need to start feeding your chicken a high-quality diet (if you do not do so already) to ensure that he looks his best on show day. On the day of the show itself you should wait to feed and water your chicken until after the judging is complete to make sure he stays clean. Do not wait too long after the judging, however, because your chicken will be very hungry and thirsty.

Ayam Cemani Chickens

Chapter Ten: Ayam Cemani Care Sheet

In reading this book you have received a wealth of information about the Ayam Cemani chicken breed. By the time you finish this book you should have a good idea whether this is the right breed for you and, if it is, you should be properly equipped to get started in raising your own Ayam Cemani chickens. As you care for your chickens, however, you may have questions or you might want to reference certain pieces of information from this book. Rather than flipping through the entire book, use this care sheet to find the most important facts about owning and caring for these beautiful chickens.

Chapter Ten: Ayam Cemani Care Sheet

1.) Basic Information

Origins: island of Java, Indonesia

Species: *Gallus gallus domesticus* (domesticated chicken)

Breed Name: Ayam Cemani

Coloration: completely black, inside and out

Genetics: hyperpigmentation caused by a dominant gene (Fibromelanosis)

Weight (male): between 2 and 3.5 kg (4.4 to 7.7 lbs.)

Weight (female): between 1.5 and 2.5 kg (3.3 to 5.5 lbs.)

Beak (length): 1.8 cm (0.7 inches)

Beak (width): between 0.9 and 1.2 cm (0.35 to 0.47 inches)

Comb: single, black; 3, 5 or 7 points

Wattle: couple, smooth, black

Body (male): slants back from the neck; back is thin

Body (female): rectangular, rhombus-like in shape; back is flat and wide

Tail: scooped like a horse's tail

Egg Laying: about 60 to 100 eggs during the first year; 20 to 30 eggs per cycle

Breeding Cycle: hens take a 3 to 6 month break after each cycle

Chapter Ten: Ayam Cemani Care Sheet

Egg Size: about 45 grams at laying; large compared to the size of the hen

Egg Color: cream-colored

2.) Habitat Set-up Guide

Habitat Options: coop, fenced-in enclosure, free-range

Minimum Floor Space: at least 3 square feet (90 sq. cm) per chicken

Minimum Perch Space: at least 12 inches (30 cm) per bird

Nest Boxes: one per four hens, minimum of two

Nest Box Lining: straw, not hay

Coop Materials: plastic, recycled materials, wood

Coop Types: ark, chicken house, poultry shed, portable coop

Plastic Coop: durable, easy to clean, versatile; may come with a small run, needs to be used on flat ground

Eco Coop: made from recycled materials, eco-friendly option; easy shipping and transport, easy to clean, option for customized run

Wooden Coop: very cost-efficient, wide variety of design options, can be used with custom run

Chapter Ten: Ayam Cemani Care Sheet

Chicken Ark: triangular or rectangular design; keeps chickens completely contained; recommended for flocks under 12 birds

Traditional Chicken House: large wooden structure with integrated run; great DIY option

Poultry Shed: simple design, easy to DIY, pairs well with a portable run; can be built on skids for portability

Portable Coop: great for modified free-range option; provides a safe haven for free-range chickens; can be built on skids or wheels for transport

Factors to Consider for Urban Areas: city ordinances, neighbors, stray dogs, varmints, vandalism

Precautions to Take: blend coop design with existing buildings; use a sturdy fence; install security lights; plant shrubs around the enclosure to camouflage and deaden noise

Maintenance Tasks: replace bedding once a week; clean out coop and nesting boxes; use a poultry detergent like Poultry Shield

Keeping Chickens Warm: keep them in an enclosure during winter in cold areas; ensure protection from drafts and proper ventilation

Chapter Ten: Ayam Cemani Care Sheet

3.) Nutritional Information

Nutrients Needed: protein, carbohydrate, fat, vitamins, minerals, water

Amino Acids: arginine, lysine, methionine, and threonine

Protein Needs: highest during development; raises again during laying

Protein (chicks): 18% until week 6, 16% until week 12, 15% until week 18

Protein (laying hens): about 18.8%

Water: 2 to 3 times feed consumption by weight

Feed Types: starter, grower, layer/breeder, medicated

Starter Feed: 18% to 22%protein, , for chicks up to 6 weeks old

Grower Feed: 15% to 16%protein, , for chicks up to 14 weeks

Developer Feed: 14% to 15% protein, for chicks up to 20 weeks

Layer Feed: about 16% protein, extra calcium, start at 20 weeks

Breeder Feed: for breeding hens, supplemented with extra vitamins and protein

Medicated Feed: medicated to prevent common disease

Chapter Ten: Ayam Cemani Care Sheet

Other Foods: scratch grains, grit, vegetables, fruit, table scraps

Feeding Recommendations: free-choice until 14 to 18 weeks then switch to adult feed

Foods to Avoid/Limit: cracked corn

4.) Breeding Tips

Breeding Type: polygynous

Male Chicken: rooster or cock

Female Chicken: hen

Eggs per Year: about 60

Egg Size: about 45g

Egg Hatching: must be artificially incubated

Courting Behavior: male dips one wing and circles the female; hen crouches to indicate receptiveness

Breeding Behavior: male mounts the hen and treads on her back

Fertilization: can only occur during 15 minutes of the egg's journey through the female's system; female can be fertile for 2 weeks or more

Egg Laying: egg travels through the female's system for 25 hours

Chapter Ten: Ayam Cemani Care Sheet

Storing Eggs: maximum 10 to 14 days; tilt or turn once daily; store pointy-end down

Storage Temperature: 55° to 60°F (13° to 15.5° C)

Storage Humidity: between 70% and 75%

Preparing the Incubator: run it for 24 hours (minimum) before adding eggs

Incubation Temperature: 37.5°C (99.5°F)

Incubation Humidity: 50% for the first 18 days and then 55% to 60% for the last 3

Turning Eggs: daily for first 18 days, none during last 3

Incubation Period: 21 days average

Chapter Eleven: Relevant Websites

Raising Ayam Cemani chickens can be a challenge, especially if you have never kept poultry before. To help you get started on the right foot you will find a collection of relevant websites in this chapter. Here you will find resources to help you choose the right food for your chickens and to build, buy, or stock your chicken coop. You will also find resources for breeding and incubation supplies as well as general resources for caring for backyard chickens.

Chapter Eleven: Relevant Websites

1.) Food for Ayam Cemani Chickens

The key to keeping your Ayam Cemani chickens healthy is to provide them with a nutritious diet. Below you will find an assortment of relevant websites to help you feed and water your chickens:

United States Websites:

"Natural Chicken Feed." Nutrena.

www.nutrenaworld.com

"Organic Chicken Feed." Scratch and Peck Feeds.

www.scratchandpeck.com

"Chicken Feed and Treats." Tractor Supply Co.

www.tractorsupply.com

"Homemade Chicken Feeder & Waterer Designs."
Backyard Chickens.

www.backyardchickens.com

Chapter Eleven: Relevant Websites

United Kingdom Websites:

"Poultry Feeds, Treats & Grit." Flyte So Fancy.

www.flytesofancy.co.uk

"Hi Peak Organic Chicken Feed." Hens for Pets.

www.hensforpets.co.uk

"Organic Poultry Feeds." The Organic Feed Company.

www.organicfeed.co.uk

"Feeders & Drinkers." Chicken Hill.

www.chickenhill.co.uk

"Feeding Chickens." Keeping-Chickens.me.uk

www.keeping-chickens.me.uk

Chapter Eleven: Relevant Websites

2.) Coops and Coop Supplies

To keep your Ayam Cemani chickens safe from predators and protected from inclement weather you need a chicken coop. Below you will find an assortment of relevant websites to help you build, buy or stock your chicken coop:

United States Websites:

"Chicken Coops." Petco.

www.petco.com

"Build a Chicken Coop." Tractor Supply Co.
www.tractorsupply.com

"Poultry Equipment." FarmTek.
www.farmtek.com

Chapter Eleven: Relevant Websites

United Kingdom Websites:

"Poultry Houses and Chicken Coops." Green Valley Poultry Supplies.

www.chicken-house.co.uk

"Omlet Coops." Chicken Hill.

www.chickenhill.co.uk

"Chicken Coops and Houses." Egg Shell.
www.eggshellonline.co.uk

"Chicken Houses, Poultry Supplies and Complete Packages." Chicken House Company.
www.thechickenhousecompany.co.uk

"Free Range Chicken Houses." Smiths Sectional.
www.smithssectionalbuildings.co.uk

Chapter Eleven: Relevant Websites

3.) Breeding and Incubation Supplies

Breeding Ayam Cemani chickens can be a challenge but it can also be very rewarding. Below you will find an assortment of relevant websites to help you breed your chickens and to incubate/hatch the eggs:

United States Websites:

"Choosing an Incubator & Incubation Tips." My Pet Chicken.

www.mypetchicken.com

"Homemade Chicken Brooder Designs." Backyard Chickens.

www.backyardchickens.com

"Brooders and Accessories." Stromberg's Chicks & Game Birds Unlimited.

www.strombergschickens.com

"Incubators." Incubator Warehouse.com.

www.incubatorwarehouse.com

Ayam Cemani Chickens

Chapter Eleven: Relevant Websites

United Kingdom Websites:

"Incubation Accessories." Chicken House Company.

www.thechickenhousecompany.co.uk

"Brooders and Brooding." P&T Poultry.

www.pandtpoultry.co.uk

"Incubation and Rearing." Omlet.

www.omlet.co.uk

"Chicken Incubators and Accessories." Green Valley
Poultry Supplies.

www.chicken-house.co.uk

"Brooders and Chick Care." Ascott Dairy.

www.ascott-dairy.co.uk

Chapter Eleven: Relevant Websites

4.) General Info for Raising Chickens

You can never have enough information about raising and caring for Ayam Cemani chickens. Below you will find an assortment of relevant websites full of valuable information about raising and caring for chickens in general:

United States Websites:

"Raising Chickens 101." The Old Farmer's Almanac.

www.almanac.com/home-pets-family/raising-chickens-blog

"Your Guide to Raising Chickens." Chickens 101.

www.chickens101.com

"Learn to Raise Backyard Chickens." Rodale's Organic Life.

www.rodalesorganiclife.com/garden/learn-raise-backyard-chickens

Ayam Cemani Chickens   P a g e | **112**

Chapter Eleven: Relevant Websites

"Keeping Chickens." Poultry Keeper.

www.poultrykeeper.com/keeping-chickens

Chapter Eleven: Relevant Websites

United Kingdom Websites:

"Chickens as Pets." RSPCA.

www.rspca.org.uk/adviceandwelfare/farm/farmanimals/chickens

"Keeping Chickens – What You Need to Know." NI Direct.

www.nidirect.gov.uk/keeping-chickens-what-you-need-to-know

"Backyard Protein: An Urbanite's Guide to Raising Chickens." Breaking Muscle.

www.breakingmuscle.co.uk

"Keeping Chickens." Longdown Activity Farm.

www.longdownfarm.co.uk

"Keeping Chickens for Meat – Raising & Feeding Table Birds." Low Cost Living.

http://www.lowcostliving.co.uk

Index

A

amino acids	58
antibiotics	42, 80, 85, 86, 88
APA	3, 34
appearance	7, 32, 38, 46, 87
appetite	80, 84
Avian Influenza	79, 80
Avian Pox	79, 81
avocado	65
Ayam Kedu	14, 125

B

beak	3, 9, 76, 92, 94
bedding	26, 28, 29, 53, 89, 97, 101
behavior	24, 38, 63, 70
black	1, 6, 7, 9, 10, 11, 12, 14, 15, 16, 17, 92, 93, 94, 95, 99, 122, 126, 128
blood	17, 83, 87
body	3, 7, 9, 10, 70, 84, 93
bones	1, 7
Botulism	79, 82
breathing	39, 80, 85, 86, 87, 88
breed	4, 1, 2, 5, 6, 7, 8, 9, 10, 12, 14, 15, 18, 23, 32, 33, 34, 35, 36, 38, 41, 67, 68, 72, 74, 78, 91, 92, 98, 110, 124
breeder	33, 34, 36, 38, 40, 41, 61, 66, 67, 72, 102
breeder formula	61
breeding	4, 2, 8, 22, 35, 38, 66, 67, 68, 70, 102, 105
brooder	26, 27, 29, 110
brooding	3, 30, 85

C

calcium	58, 61, 64, 84, 102
cancer	87

carbohydrate	58, 65, 102
care sheet	98
chest	9, 76, 93, 94
chicken ark	46

chicken coop 2, 3, 26, 27, 29, 30, 44, 45, 46, 48, 50, 51, 52, 53, 56, 90, 105, 108

chicks 3, 26, 27, 28, 29, 33, 34, 36, 38, 39, 49, 60, 61, 66, 67, 68, 70, 74, 85, 87, 88, 102, 122

claws	9, 93, 94
cloaca	5, 83
Coccidiosis	79, 83
cockerel	4
coloration	2, 7, 9, 10, 12, 14, 15
comb	9, 14, 15, 70, 84, 92, 94
commercial feed	64
contaminated	80, 83, 85, 88, 89

coop 27, 29, 41, 42, 43, 44, 45, 46, 47, 48, 49, 50, 51, 52, 53, 54, 78, 83, 89, 90, 100, 101, 108, 122, 125

corn	63, 66, 103
costs	26, 28, 29, 30, 31, 75
courting	70
crown	1
culling	49, 67, 74, 75

D

DEFRA	20
dehydration	80
developer feed	61
diarrhea	39, 65, 80, 84
diet	42, 56, 57, 58, 60, 64, 65, 78, 79, 84, 89, 97, 106
discharge	39, 85, 86, 88
disease	19, 24, 39, 53, 54, 80, 81, 82, 83, 85, 86, 87, 88, 89, 102
dislocation	75, 76
domesticated	9, 10, 11, 15, 22, 57, 68, 70, 74, 79, 99
ducks	24

E

Egg Binding 79, 84

eggs 3, 4, 5, 8, 10, 11, 15, 16, 22, 34, 42, 44, 49, 61, 62, 67, 68, 70, 72, 73, 74, 75, 84, 85, 99, 104, 110, 122, 126

enclosure 3, 22, 28, 32, 38, 45, 46, 47, 48, 49, 54, 100, 101

eyes 9, 39, 54, 85, 86, 87, 93, 94

F

facts 4, 2, 6, 98

fat 42, 58, 66, 102

feathers 1, 3, 4, 7, 9, 10, 14, 16, 39, 70, 74, 82, 93, 94

feces 39, 83, 87, 89

feed 5, 22, 29, 51, 56, 59, 60, 61, 63, 64, 65, 66, 75, 82, 83, 85, 86, 88, 89, 97, 102, 103, 106, 107, 126

feeders 30, 47, 51, 89

female 4, 9, 11, 68, 70, 84, 99, 103

fertile 4, 22, 49, 71, 103, 126

fertilization 70, 71

flock 19, 20, 45, 48, 49, 51, 66, 68, 74, 80, 82, 85, 86, 89

food 4, 6, 24, 25, 26, 28, 29, 42, 57, 61, 63, 64, 65, 66, 75, 82, 85, 88, 105

fowl 4, 12, 25, 82, 86

free-range 15, 16, 22, 29, 42, 47, 49, 51, 100, 101

fruit 63, 65, 103

G

gene 10, 11, 15, 99

grit 63, 64, 103

grower 60, 62, 102

growth 5, 42, 59, 60, 86

guinea fowl 25

H

habitat 26

healthy	2, 38, 55, 56, 57, 58, 60, 68, 78, 79, 89, 106, 125
heat	27, 54
hen	1, 3, 4, 10, 11, 22, 69, 70, 74, 75, 84, 85, 100, 103, 128
history	2, 6, 124
humidity	72, 73
hyperpigmentation	10, 11, 99

I

immune	83, 85, 89
incubator	4, 72, 73, 110
Indonesia	1, 7, 10, 13, 14, 99, 126
Infectious Bronchitis	79, 85
Infectious Coryza	79, 86
influenza	80
initial costs	26, 28
insects	63, 80

J

Java	1, 7, 10, 12, 99

K

Kedu	14, 127
Kedu Hsian	14

L

laws	19, 20
layer formula	61
laying	4, 10, 11, 58, 62, 64, 66, 74, 75, 85, 88, 100, 102
legal	19, 20, 26, 127
lesions	81
lethargy	80
license	19
licensing	18, 19

Ayam Cemani Chickens

lifespan	14
lungs	54

M

male	3, 9, 11, 15, 24, 68, 70, 71, 99, 103
Marek's Disease	79, 87
mating	70, 71
meat	1, 2, 3, 4, 13, 17, 22, 60, 77, 93, 95
medicated feed	61
medicine	13
mites	39, 53
molt	74, 75
monthly costs	26, 30
mortality	80, 88
mosquitos	81

N

neck	9, 11, 70, 75, 76, 77, 82, 93, 94, 99
nest	24, 43, 44, 53, 84
nest box	44, 84
Newcastle Disease	79, 88
noise	21, 48, 49, 97, 101
nose	3, 39, 85, 86
nutrients	2, 57, 58, 59, 60, 89
nutrition	80, 84, 97, 113, 127
nutritional needs	56, 57, 58, 79, 89

O

oyster shell	64

P

paralysis	82, 87, 88
parasite	83

partridge	12, 14
pathogenic	80, 88
pecking order	24
pen	96, 97
pheasants	24, 86
phosphorus	58
portable	42, 44, 46, 47, 52, 100, 101, 108
poultry	3, 4, 5, 19, 20, 24, 33, 34, 36, 38, 40, 42, 43, 44, 46, 47, 53, 63, 66, 67, 74, 81, 91, 96, 97, 100, 101, 105, 106, 108, 111, 124, 126, 127
poultry show	91, 96, 97
pre-fabricated	45
price	8, 17, 67
production	23, 61, 68, 75, 80, 85, 86, 88
pros and cons	18, 32
protein	5, 58, 60, 61, 63, 66, 75, 102, 113
pullet	4

Q

quarantine	40, 89

R

relevant websites	105, 106, 108, 110, 112
repairs	28, 30
requirements	18, 19, 43, 57, 67, 127
resources	105
restrictions	19, 20, 96
roost	8, 10, 41, 43, 45
rooster	1, 3, 4, 22, 49, 71, 103

S

sanitation	80, 81, 83, 85, 86, 88, 89
scaly	39, 87
scratch grains	63, 64, 103
seeds	63

shed	46, 47, 52, 100
shelter	27, 41
showing	91, 96, 124, 126
skin	1, 3, 10, 14, 16, 81, 93, 95
space	22, 32, 42, 43, 45, 46, 47, 49, 50, 51, 52, 54, 66
species	4, 7, 8, 9, 10, 12, 14, 15, 16, 18, 24, 25, 80, 82, 88, 92
standard	3, 9, 52, 60, 68, 91, 92, 128
starter feed	29, 60, 66
supplies	105, 108
Svart Höna	15
Swedish Black Hen	10, 15, 16, 128
swelling	80, 86, 88
symptoms	80, 82, 83, 84, 85, 86, 87, 88

T

table scraps	63, 65, 103
tail	9, 11, 70, 99
temperature	3, 72, 73, 85
tongue	9, 92, 94
train	96
treatment	40, 80, 81, 82, 83, 85, 86, 87, 88
tumors	87
turkeys	24

U

urban	19, 22, 48, 49

V

vaccination	30, 81
vegetables	63, 64, 65, 103
ventilation	50, 54, 101
vet	30, 84
veterinary	26, 28, 30, 40
virus	80, 81, 85, 87

Ayam Cemani Chickens

W

water	25, 59, 65, 66, 73, 82, 85, 86, 88, 97, 102, 106
waterers	24, 30, 47, 51, 66, 89
wattle	9, 15
weather	27, 42, 108
wings	9, 70, 82, 93, 95
wingspan	9, 95

Photo Credits

Cover Photo "Orlando" by Fergus Hall

@ www.peggyspekins.co.uk

Page 1 Photo By Kangwira

via Wikimedia Commons,

Page 6 Photo By Fergus Hall

@ www.peggyspekins.co.uk

Page 18 Photo By

www.401kcalculator.org

Page 33 Photo By Minseong Kim

via Wikimedia Commons,

Page 41 Photo By ChickenMan

via Wikimedia Commons,

Page 56 Photo

By Pixabay User PublicDomainPictures,

Page 67 Photo By Fergus Hall

@ www.peggyspekins.co.uk

Page 78 Photo By Fergus Hall

@ www.peggyspekins.co.uk

Page 91 Photo By Fergus Hall

@ www.peggyspekins.co.uk

Page 100 Photo By Jeffrey Pamungkas

via Wikimedia Commons

Page 107 Photo By Fergus Hall

@ www.peggyspekins.co.uk

References

"A Beginner's Guide to Showing Poultry."

Scottish smallholder & Grower Festival.

www.ssgf.uk/exhibitors/beginners-guide-to-showing-poultry/

"Ayam Cemani." Feathersite.com.

www.feathersite.com/Poultry/CGA/Cemani/BRKCemani.ml

"Ayam Cemani." Greenfire Farms.

www.greenfirefarms.com/chicken/ayam-cemani/

"Ayam Cemani: A Rare Chicken Breed That is Black Inside Out." Amusing Planet.

www.amusingplanet.com/2014/08/ayam-cemani-rare-chicken-breed-that-is.html

"Ayam Cemani for Consumption." Cemani Farms.

www.cemanifarms.com/2013/09/ayam-cemani-for-consumption.html

"Ayam Cemani History." Cemani Farms.

www.cemanifarms.com/2013/09/ayam-cemani-history.html

"Ayam Kedu Chicken." RightPet.com.

www.rightpet.com/livestock-poultrydetail/ayam-kedu-chicken

"Chicken Coops: Choose the Right One." HobbyFarms.com.

www.hobbyfarms.com/livestock-and-pets/chicken-coop-design.aspx

"Chicken Economics: The Cost of Keeping Chickens." Mother Earth News.

www.motherearthnews.com/homesteading-and-livestock/chicken-economics.aspx

"Chicken Houses/Coops." Keeping-Chickens.me.uk.

www.keeping-chickens.me.uk/getting-started/chicken-houses

"Choosing Healthy Chickens." Keeping-Chickens.me.uk.

www.keeping-chickens.me.uk/chickens/choosing-healthy-chickens

"Common Poultry Diseases." University of Florida IFAS Extension.

www.edis.ifas.ufl.edu/ps044

"Cost of Raising Chickens." City Girl Farming.

www.citygirlfarming.com

"Feeding Chickens." PoultryKeeper.com.

www.poultrykeeper.com/keeping-chickens-faq/feeding-chickens-what-feed-chickens

"Glossary of Poultry Terms." eFowl.com.

www.efowl.com/articles.asp?id=246

Harfenist, Ethan. "Indonesia's Jet-Black Chickens are the Dark Side of Poultry." Munchies.

http://munchies.vice.com/articles/indonesias-jet-black-chickens-are-the-dark-side-of-poultry

"How to Dispatch a Chicken." PoultryKeeper.com.

www.poultrykeeper.com/general-chickens/how-to-kill-a-chicken

"How to Feed Your Laying and Breeding Hens." Oregon State University.

http://ir.library.oregonstate.edu/xmlui/bitstream/handle/1957/17469/pnw477.pdf?sequence=1

"How to Hatching Fertile Eggs." Cemani Farms.

www.cemanifarms.com/2014/01/how-to-hatching-fertile-eggs.html

"How to Prepare Your Chickens for a Poultry Show." The Guardian.

www.theguardian.com/lifeandstyle/gardening-blog/2013/nov/11/chickens-showing

"Kedu." Feathersite.com.

www.feathersite.com/Poultry/CGK/Kedu/BRKKeduEdited.html

"Legal Protections for Farm Animals." ASPCA.org.

www.aspca.org/fight-cruelty/farm-animal-cruelty/legal-protections-farm-animals

Libal, Angela. "Compatible Birds to Keep with Chickens." Pets on Mom.me.

www.animals.mom.me/compatible-birds-keep-chickens-9486.html

Ayam Cemani Chickens

Marx, Rebecca Flint. "Meet the $2,500 Chicken." Food and Wine.

www.foodandwine.com/blogs/2013/08/12/
meet-the-2500-dollar-chicken

"Nutritional Requirements of Poultry." The Merck Veterinary Manual.

www.merckvetmanual.com/
mvm/poultry/nutrition_and_management_poultry/nutri
tional_requirements_of_poultry.html

Pals, Bart. "Helping Poultry Breeders Raise Birds in an Urban Area." American Poultry Association.

www.amerpoultryassn.com/Poultry%20Articles/Helpin
g%20Poultry%20Breeders%20Raise%20Birds%20in%20a
n%20Urban%20Area.pdf

"Pelung Chickens." Cemani Farms.

www.cemanifarms.com/2013/09/pelung-chickens.html

"Rules and Regulations." Keeping Chickens: A Beginner's Guide.

www.keeping-chickens.me.uk/getting-started/rules-
and-regulations

"Standard of the Ayam Cemani." Cemani Farms.

www.cemanifarms.com/2013/09/standard-of-ayam-cemani.html

"Swedish Black Hen." Greenfire Farms.

www.greenfirefarms.com/chicken/swedish-black-hen

"Tips for Keeping Backyard Hens." Frugally Sustainable.

http://frugallysustainable.com/2012/03/tips-for-keeping-backyard-hens

"What to Look for When Buying." The Chicken Vet.

www.chickenvet.co.uk/health-and-common-diseases/what-to-look-for-when-buying/index.aspx

Printed in May 2021
by Rotomail Italia S.p.A., Vignate (MI) - Italy